STATISTICAL GEOINFORMATICS FOR HUMAN ENVIRONMENT INTERFACE

CHAPMAN & HALL/CRC
APPLIED ENVIRONMENTAL STATISTICS

Series Editor
Richard Smith
University of North Carolina
U.S.A.

Published Titles

Michael E. Ginevan and Douglas E. Splitstone, **Statistical Tools for Environmental Quality**

Timothy G. Gregoire and Harry T. Valentine, **Sampling Strategies for Natural Resources and the Environment**

Daniel Mandallaz, **Sampling Techniques for Forest Inventory**

Bryan F. J. Manly, **Statistics for Environmental Science and Management, Second Edition**

Steven P. Millard and Nagaraj K. Neerchal, **Environmental Statistics with S Plus**

Song S. Qian, **Environmental and Ecological Statistics with R**

Wayne L. Myers and Ganapati P. Patil, **Statistical Geoinformatics for Human Environment Interface**

CHAPMAN & HALL/CRC
APPLIED ENVIRONMENTAL STATISTICS

STATISTICAL GEOINFORMATICS FOR HUMAN ENVIRONMENT INTERFACE

WAYNE L. MYERS • GANAPATI P. PATIL

CRC Press
Taylor & Francis Group
Boca Raton London New York

CRC Press is an imprint of the
Taylor & Francis Group, an **informa** business

A CHAPMAN & HALL BOOK

CRC Press
Taylor & Francis Group
6000 Broken Sound Parkway NW, Suite 300
Boca Raton, FL 33487-2742

First issued in paperback 2018

© 2013 by Taylor & Francis Group, LLC
CRC Press is an imprint of Taylor & Francis Group, an Informa business

No claim to original U.S. Government works

ISBN-13: 978-1-4200-8287-6 (hbk)
ISBN-13: 978-1-138-37272-6 (pbk)

Library of Congress Cataloging-in-Publication Data

Myers, Wayne L., 1942-
 Statistical geoinformatics for human environment interface / Wayne L. Myers and Ganapati Patil.
  p. cm. -- (Chapman & Hall/CRC applied environmental statistics)
 Includes bibliographical references and index.
 ISBN 978-1-4200-8287-6 (hardcover : alk. paper)
 1. Human ecology--Statistical methods. 2. Human ecology--Mathematical models. 3. Human beings--Effect of the environment on--Statistical methods. 4. Human beings--Effect of the environment on--Mathematical models. 5. Nature--Effect of human beings on--Statistical methods. 6. Nature--Effect of human beings on--Mathematical models. I. Patil, Ganapati P. II. Title.

GF23.M35S82 2012
304.201'5195--dc23   2012014400

Visit the Taylor & Francis Web site at
http://www.taylorandfrancis.com

and the CRC Press Web site at
http://www.crcpress.com

Contents

Preface...ix
Authors...xi

1. **Statistical Geoinformatics of Human Linkage with Environment**.....1
   1.1   Introduction...1
   1.2   Human Environment Informational Interface and Its Indicators...1
   1.3   The "-matics" of Geoinformatics..3
   1.4   Spatial Synthesis of Disparate Data by Localization as
         Vicinity Variates..4
   1.5   Spatial Posting of Tabulations (SPOTing).....................................5
   1.6   Exemplifying County Context...6
   1.7   Posting Points and Provisional Proximity Perimeters for
         Lackawanna County...7
   1.8   Surveillance with Software Sentinels..10
   1.9   Backdrop: Distributed Data Depots and Digital Delivery...........12
   References...13

2. **Localizing Fixed-Form Features**...15
   2.1   Introduction...15
   2.2   Locality Layer as Poly-Place Purview...15
   2.3   Localizing Layer of Proximity Perimeters...................................19
   2.4   Localizing Linears by Determining Densities..............................22
   2.5   Transfer from Perimeters to Points...24
   2.6   Apportioning Attributes of Partial Polygons...............................28
   2.7   Backdrop: GIS Generics...28
   References...30

3. **Precedence and Patterns of Propensity**..31
   3.1   Introduction...31
   3.2   Prescribing Precedence...34
   3.3   Product–Order Precedence Protocol...35
   3.4   Precedence Plot..36
   3.5   Propensities as Progression of Precedence..................................38
   3.6   Progression Plot...40
   3.7   Reversing Ranks...41
   3.8   Inconsistency Indicator...42
   3.9   Backdrop: Statistical Software...43
   References...45

4. Raster-Referenced Cellular Codings and Map Modeling 47
   4.1   Introduction 47
   4.2   Fixed-Frame Micromapping with Conceptual Cells 47
   4.3   Cover Classes and Localizing Logic 48
   4.4   Raster Regions and Associated Attributes 51
   4.5   Map Modeling 52
   4.6   Layer Logic 55
   References 56

5. Similar Settings as Clustered Components 57
   5.1   Introduction 57
   5.2   CLAN Clusters 59
   5.3   CLUMP Clusters 64
   5.4   CLAN Cluster Centroids... 68
   5.5   Salient Centroids 69
   5.6   Graded Groups by Representative Ranks 71
   5.7   Rank Rods 72
   5.8   Salient Sequences by Representative Ranks 74
   References 78

6. Intensity Images and Map Multimodels 79
   6.1   Introduction 79
   6.2   Intensity as Frequency of Occurrence.. 79
   6.3   Hillshades and Slopes 86
   6.4   Interposed Distance Indicators 88
   6.5   Backdrop: Pictures as Pixels and Remote Sensing........................ 90
   References 93

7. High Spots, Hot Spots, and Scan Statistics 95
   7.1   Introduction 95
   7.2   SaTScan™ 96
   7.3   Concentration of CIT Core Development 97
   7.4   Complexion of CIT Developments 100
   7.5   Particular Proximity 104
   7.6   Upper Level Set (ULS) Scanning 109
   7.7   Backdrop: Python Programming 110
   References 110

8. Shape, Support, and Partial Polygons 113
   8.1   Introduction 113
   8.2   Inscribed Octagons 113
   8.3   Matching Margins and Adjusting Areas.................................... 117
   8.4   Shape and Support for Local Roads 119
   8.5   Precedence Plot for Shapes and Supports 121

8.6   Supports Spanning Several Partial Polygons 123
References 124

9. **Semisynchronous Signals and Variant Vicinities** 125
   9.1   Introduction 125
   9.2   Distal Data 129
   9.3   Median Models 132
   9.4   Pairing/Placement Patterns of Signal Strengths 136
   Reference 141

10. **Auto-Association: Local Likeness and Distance Decline** 143
   10.1   Introduction 143
   10.2   Cluster Companions 143
   10.3   Kindred Clusters 147
   10.4   Local Averages 147
   10.5   LISA: Local Indicator of Spatial Association 151
   10.6   Picking Pairs at Lagged Locations 152
   10.7   Empirical (Semi-)Variogram 155
   10.8   Moran's *I* and Similar Spatial Statistics 158
   References 160

11. **Regression Relations for Spatial Stations** 161
   11.1   Introduction 161
   11.2   Trend Surfaces 161
   11.3   Regression Relations among Vicinity Variates 164
   11.4   Restricted Regression 171
   References 171

12. **Spatial Stations as Surface Samples** 173
   12.1   Introduction 173
   12.2   Interpolating Intensity Indicators as Smooth Surfaces 177
   12.3   Spline Smoothing 180
   12.4   Kriging 181
   References 182

13. **Shifting Spatial Structure** 183
   13.1   Introduction 183
   13.2   Space–Time Hotspots 183
   13.3   Salient Shifts 183
   13.4   Paired Plots 186
   13.5   Primary Partition Plots 186
   13.6   Backdrop: Spectral Detection of Change with Remote Sensing ... 189
   References 189

14. Synthesis and Synopsis with Allegheny Application 191
   14.1 Introduction 191
   14.2 Localization Logic... 191
   14.3 Locality Layer.. 193
   14.4 Localizing Layer.. 194
   14.5 Poly-Place Purviews... 196
   14.6 Significant Spatial Sectors with Scan Statistics............................ 197
   14.7 Scale Sensitivity and Partial Precedence 198
   14.8 Cluster Components and Cluster Companions............................. 200
   14.9 Trend Surfaces 202
   14.10 Surveillance Systems: Sentinel Stations and Signaling.............. 205
   14.11 Scripted Sentinels... 206
   14.12 Smart-Sentinel Socialization 206
   References 207

Index 209

Preface

We deliberately depart from conventional concepts of both space and inter-face with regard to human/environment. Major benefits of our approach are (1) capacity to cope with complexity, (2) systematic surveillance, (3) visual-ization and communication, (4) preliminary prioritization, (5) coupling of geographic information systems and statistical software, and (6) avenues for automation. We treat space as a pattern of proximities or vicinities, with the pattern being a square grid and the vicinities being centrically referenced with regard to placement. The thematic targets are thus *localized*, and the pro-cess of *localization* is one of two paradigms that we present. Both the spacing of placement for the grid and the scopes of vicinities can be varied continu-ously as aspects of analysis. Differences in the duality of center spacing and proximity purview (spatial scope) lend a multiscale aspect to analysis, with proximity purview corresponding to the concept of support in geostatistics. Double grids with offset centers constitute one among many specific scaling strategies. Speed is one analogy for effects of changing support, with local variations increasingly masked as speed is averaged over greater distance.

We treat human/environment interface as an abstraction to be explored through *indicators* rather than being subject to an all-encompassing arguable definition. Thus, the multitudinous aspects of such interface are reflected in multiple indicators, which is the second paradigm that we present. An indicator is integrative over a vicinity, and quantification is thus as *integrative vicinity indicators* constituting statistical *vicinity variates*. The set of vicinity variates that is localized on a common centric reference position (or posting point) thus constitutes an observation or case in the statistical sense. This approach neutralizes the common conundrum of how to reconcile disparate spatial structures such as points, lines, and polygons. We make no compen-dium of indicators but rather choose those that serve for procedural explica-tion of our approach.

Due to this alternate adaptive approach of paired paradigms, we also depart from convention in pattern of presentation. Reading the first and last chapters should convey the concepts of localization and indication that are the essence of the approach. Strategies of statistical and spatial analysis are introduced in the intervening chapters, which should be read in order because each builds upon those preceding. Thus, the organization is some-thing of a sandwich, with the meat in the middle and extremities being the basis for grasping it.

The tactical and technical tools needed to apply our approach are dis-tributed over different disciplinary domains. Acquiring full functional facility with all aspects can consume a lengthy career and requires multi-disciplinary modality. Both of us have had lengthy careers extending into emeritus status, and we have worked in many multidisciplinary modes as well as closely cooperating. These transdisciplinary tracts often entail tangled terminologies. We do some accommodation of diversity in disciplines through chapter "backdrops." These come where appropriate application would require some depth in a discipline. A backdrop is meant to be a portal into a disciplinary purview rather than a tutorial. Those who already have considerable capability in that discipline are likely to dismiss the backdrop as being excessively elementary. Accordingly, we simply invite them to skip it and move on. However, the pointers in the backdrop are the kinds of leads we would have appreciated and often did not have when doing self-study. Likewise, we make no apology for ignoring some conflicting terminology. We choose terms and invent acronyms while endeavoring to explain what we mean. Beyond that, different disciplinarians are expected to do their own transliteration.

We acknowledge the roles of many institutional and individual influences in shaping our scenarios. Institutional influences include the National Science Foundation (NSF), U.S. Geological Survey (USGS), U.S. Agency for International Development (USAID), NASA, NOAA, Penn State University, and Michigan State University, along with several international involvements. Our orientation is toward intellectual inquiry and exploration rather than establishing canons of computation. We dedicate this to pluralism, perseverance, and patience, with particular thanks to our spouses for their patient support and to institutional administrators for their tolerance of our idiosyncrasies.

Authors

Wayne L. Myers is Professor Emeritus of Forest Biometrics at the Pennsylvania State University. He specializes in landscape analysis using geographic information systems (GIS) and remote sensing in conjunction with multivariate approaches to analysis and prioritization. He is a Certified Forester under the auspices of the Society of American Foresters, an Emeritus Member of the American Society of Photogrammetry and Remote Sensing, and a 40-year member of the American Statistical Association.

Dr. Myers earned bachelor's, master's, and PhD degrees from the University of Michigan. He served as a regional biometrician in Canada and then on the faculty of Michigan State University before moving to Penn State University in 1978. At PSU he joined the faculties of the Intercollege Graduate Degree Programs in both Ecology and Operations Research.

Dr. Myers served a two-year assignment as Forestry Advisor to USAID/India Mission under a Joint Career Corps contract between PSU and USAID during 1988 and 1989, and he has also worked in Malaysia along with other international settings. Before formal retirement in 2009, he served as Assistant Director for Graduate Studies in the School of Forest Resources at PSU and as Director of the Office for Remote Sensing and Spatial Information Resources in the Penn State Institutes of Energy and Environment. He has written several books and/or parts of books along with numerous research articles.

Ganapati P. Patil is Director of the Center for Statistical Ecology and Environmental Statistics and Distinguished Professor Emeritus of Mathematical and Environmental Statistics at the Pennsylvania State University. He is a fellow of the American Statistical Association, American Association of Advancement of Science, Institute of Mathematical Statistics, International Statistical Institute, Royal Statistical Society, International Association for Ecology, International Indian Statistical Association, Indian National Institute of Ecology, and Indian Society for Medical Statistics.

Dr. Patil has served on panels for numerous organizations, including the United Nations Environment Program, U.S. National Science Foundation, U.S. Environmental Protection Agency, U.S. Forest Service, and U.S. National Marine Fisheries Service. He has authored/coauthored more than 300 research papers and more than 30 cross-disciplinary monographs, and he has been the founding Editor-in-Chief of the international journal, *Environmental and Ecological Statistics*.

1

Statistical Geoinformatics of Human Linkage with Environment

1.1 Introduction

For present purposes, *Statistical Geoinformatics for Human Environment Interface* is concerned with spatial patterns of interaction between complexes of environmental process factors and human process factors. The process aspect is very important because it underscores the temporally dynamic nature of the phenomena of interest. The phrasing in terms of complexes of factors is equally important because it casts the suites of concerns in a multivariate context. Continuously varying scales of both time and space are interwoven throughout these concerns. This book occupies multiple middle grounds between geographic information systems (GIS), multivariate analysis, partial ordering, spatial statistics, and even remote sensing. At the interface of several such disciplinary domains, it becomes a challenge in itself to avoid a tangle of terminologies. Inasmuch as we do not intend to reiterate any of these disciplinary domains directly, we will often choose to use terminology that is not directly derived from a discipline but clarifies our context sufficiently in a way that each specialist can make whatever translation of terms they deem necessary to satisfy their strictures. Rather than directly diversifying a domain, we will provide synergistic strategies for a supporting science of surveillance through an innovative paradigm of *localizing* spatial data that GIS treat as layers. The book is written such that the first and last chapters can be read ahead to obtain a basic appreciation of the intention and approach that is the main thrust throughout the book. However, the other chapters should be read in numerical order thereafter because each builds upon those preceding.

1.2 Human Environment Informational Interface and Its Indicators

Human environment information interface is bidirectional with respect to obtaining information about environmental characteristics that are relevant to human habitation and also concerning how humans influence the earth environments that support them. There is a not-so-recent book entitled *Environomics of Environmentally Safe Prosperity* (Khavari 1993) that explores ways of attaining general economic prosperity while improving or at least sustaining environmental conditions. A fully functional human environment information interface is crucial to that goal, but the human side will in any case be concerned with the aspects of environment that are important to satisfying real or perceived needs of resource acquisition, access, and site suitability among others. Our heading for this side of our subject will be *human habitation*. The complementary concerns are cultural conditioning of ecological and environmental elements and the stewardship of sustainability. Our heading for this "other" side of externalities will be *earth environment and ecosystems* (Forman 2008; Fortin and Dale 2005). There is an extensive and expanding literature on the informational infrastructure of landscape ecology that could be called geospatial ecoinformatics, and Websites such as www.ecoinformatics.org in that terminological track also appear. In fact, our subject is so fluid that Web work must be considered a crucial complement to more traditional publication. With regard to references, we tend to be parsimonious by citing what we consider as references to references for a particular disciplinary domain.

Having said the foregoing to set some structure, it must be acknowledged that *human environment interface*, whether informational or otherwise, remains an abstraction that cannot be completely captured in any single simple scenario. Thus, we enter an *age of indicators* (Myers and Patil 2006) whereby various aspects of interactions at the interface are addressed by observing purported effects and/or bellwethers in the manner of caged canaries in coalmines, with these then being considered as a complex of criteria. The 1993 convention of 12 countries in Montreal, Canada, giving rise to the "Montreal Process" of identifying a framework of criteria and indicators for tracking progress in forest sustainability may be considered as a mile marker for this age of indicators. The consequent criteria or informational indicators can cover a spectrum of spatial scales and scopes that are regionally relevant. To be operationally oriented, however, such indicators must have some localization. As but one illustration, localization in a Montreal mode is reflected in a *Neighborhood Sustainability Indicators Report on a Best Practice Workshop* by Joza and Brown (2005). It is noteworthy that the concluding days of 2011 bring an expert consultation in New Delhi, India, on "Statistical Methodologies for Developing Composite Index for Environmental Sustainability" hosted by the Society for Development Studies and supported by United Nations Environment Programme (UNEP).

A suite of indicators for a situational setting must have particular polarity, specific scale and scope, complementarities in the context and not have redundancy to the degree that would constitute "stacking the deck" from a particular perspective. Indicators can be simpler surrogates for complex conceptual constructs, systematically synthesized surrogates, or even suggestive surrogates obtained from somewhat subjective "soft" data (Luloff et al. 2011). It must be anticipated that multiple indicators will not always carry complete consensus in regard to a particular perspective, which may permit only partial ordering in a multivariate mode (Bruggemann and Patil 2011). Our primary purpose is synthesizing suites of spatially specified indicators and seeking salient structure among such spatial settings by mining major messages, muting minor messages, and procuring preliminary prioritization. We will assume that the informational ores reside mostly in spatially specific GIS databases (Campagna 2006; DeMers 2009; Maguire et al. 2005; Obermeyer and Pinto 2008; Panigrahi 2010) and are to be extracted from GIS for refinement in auxiliary statistical software systems without sacrificing spatial specificity. Thus, our threefold foci are *localization*, *indication*, and *prioritization*.

1.3 The "-matics" of Geoinformatics

The populist Web reference Wikipedia is an interesting starting point in this regard, which begins by stating that a GIS is a system designed to capture, store, manipulate, analyze, manage, and present all types of geographically referenced data. This comes with a referral to the Environmental Systems Research Institute (ESRI) along with links to several source articles and Websites with various definitional and introductory perspectives. A more heavily human perspective comes from Ehlers (2008) as "the art, science or technology dealing with the acquisition, storage, processing, production, presentation and dissemination of geoinformation," with "art" interestingly in the beginning. An essential element here is "georeferencing," whereby each item of information is explicitly placed or placeable with regard to position on the earth, albeit often in different positional reference systems and with varying degrees of location precision. The geoinformatic umbrella includes specific technologies of spectral remote sensing and global positioning systems (GPS). From the term itself, it would seem that seismic information should likewise be included; however, this is conventionally not the case.

It is implicitly evident that the term *geoinformatics* is indirectly indicative of automation but is not primarily so, as also mathematics does not primarily mean "automathics." However, the term is quite suggestive of automating acquisition and utilization of georeferenced information—and the nontechnical person seems rather invited to pick up that flavor. This is especially true in view of everyday acquaintance with the very impressive navigational GPS for automobiles that incorporates roadmaps of an entire country with automated routing capability and ancillary information in a package for which miniaturization is limited largely by readability of the screen. If there were not some intent to underscore automation, then the more venerable designation of GIS probably would not have needed a "showy" surrogate of geoinformatics. Our contributions will focus on disentangling the typical topologies of georeferenced databases in order to synthesize spatially specified indicators that are amenable to export for exploration in more general statistical software systems.

Both GIS and statistical software systems give us some presentational predicaments inasmuch as a reader may be quite familiar with one and much less so for the other. Our compromise is to include a backdrop as an end element in several chapters. This chapter has a backdrop of distributed data depots and digital delivery. A reader feeling familiar with the context of a backdrop can simply bypass it. References in the backdrops are particularly selected as "references to references."

1.4 Spatial Synthesis of Disparate Data by Localization as Vicinity Variates

Geoinformatics and its pseudonyms have been and continue to be characterized by deep diversity of data domains. Informally, the major variants involve points, lines, patches, pictures (imagery), and verticality or lack thereof. Furthermore, each major genre has variants regarding schemas for representation on digital media. It is virtually inconceivable that any book carrying one of these geoinformatic monikers could avoid dealing with these disparities in data. So much so that this is addressed in the second chapter of a book on research agenda for geographic information sciences (McMaster and Usery 2004) that is highlighted with updates by the University Consortium on Geographic Information Sciences (UCGIS; www.ucgis.org) as "spatial data acquisition and integration."

For us here, the *integration* aspect becomes something of an organizing framework for the book as we consider in subsequent chapters different sorts of data and which most often yield what kinds of information on the several dimensions of the human environment interface. Furthermore, issues of integration provide the major motivations for innovative approaches on which we elaborate in the next sections. Our thematic thrust will entail commonality of context in a spatial setting such that multivariate modes are manageable. This is possible primarily through *localization as vicinity variates*.

Localization begins by creating a *localities layer* of *posting points* arranged in a grid-posting pattern. A *proximity protocol* is then established that sets the vicinity in which a variate is to be compiled around each posting point, thus giving rise to the *vicinity variate* terminology. The set of vicinities constitutes a *localizing layer*. There are two major variants of vicinity variates. An *integrative vicinity indicator* (IVI) is compiled within a particular perimeter or polygon around the respective posting point. An *interposed distance indicator* (IDI) measures the distance from a posting point to an instance or occurrence of something of interest. IVIs and IDIs are thus proximate properties of posting points. Our interest can be stated somewhat more simply as determining and analyzing proximate properties of posting points in localizing layers. Different vicinity variates posted on the same set of points comprise a multivariate dataset. The strength of quantification for each variate must be respected in the course of analysis along with statistics of spatial auto-associative aspects.

1.5 Spatial Posting of Tabulations (SPOTing)

We can portray our procedural primaries as concentric triangles in a six-pointed SPOTing star. Figure 1.1 has an upper (heavy) triangle for obtaining vicinity variates with a lower (light) triangle for multivariate matrices. Posting points are generated in *a* of the upper triangle. Proximities of posting points are set in *b* of the upper triangle. Properties as values of the vicinity variate are produced for posting point proximities in *c* of the upper triangle. The values of this vicinity variate comprise a layer in a GIS database. In order to have a multivariate dataset, the layers for different vicinity variates must be combined. This is accomplished by matching the several sets of vicinity

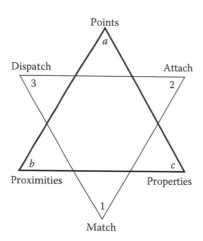

FIGURE 1.1
Twin-triangle SPOTing star schematic for generating vicinity variates as multivariate matrix (see text for details).

values to the posting points (1) and attaching them as additional attributes (2) in the lower triangle. The resulting multivariate point layer can then be dispatched for analysis in an external statistical system (3) or analyzed within the GIS itself.

The pairing of opposite corners $c3$ in Figure 1.1 represents the culmination of the localization process for transport to a statistical software system. The pairing of complementary corners $c2$ will suffice for analysis to be conducted using facilities of the parent GIS. GIS analysis facilities tend to be geared more to univariate work than to multivariate methods.

1.6 Exemplifying County Context

We use counties in the State of Pennsylvania, U.S.A., to exemplify our approaches. Figure 1.2 shows county outlines and courses of major rivers to provide a broad perspective. Three of the counties are labeled in Figure 1.2, of which Lackawanna County will be our major exemplar. Neighboring Wyoming County and Luzerne County are added later in Chapter 12 to give a larger area for particular purposes.

Lackawanna contains the urban area of Scranton, and Luzerne contains the companion city of Wilkes-Barre, whereas Wyoming is more pervasively

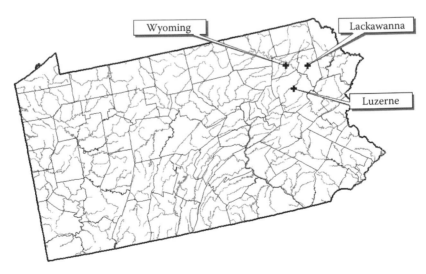

FIGURE 1.2
State of Pennsylvania showing county boundaries and major rivers as context, with names for a tricounty area.

rural. Scranton and Wilkes-Barre together occupy much of the elongate Wyoming Valley situated in an otherwise rugged region.

1.7 Posting Points and Provisional Proximity Perimeters for Lackawanna County

Figure 1.3 shows a 5-km pattern of posting points for Lackawanna County constituting a locality layer.

Figure 1.4 shows provisional proximity perimeters having a 2-km radius around the 5-km posting points in Lackawanna County. These proximity perimeters are obtained by a GIS buffering operation. These buffer perimeters are a provisional intermediate step toward a localizing layer for vicinity variates inasmuch as some of the perimeters extend beyond the boundaries of the county. Further details of valid vicinities are given in the subsequent chapter on localization of fixed-form features.

It is important to purvey a sense of the terrain for Lackawanna County in order to better appreciate the degree to which results in subsequent chapters

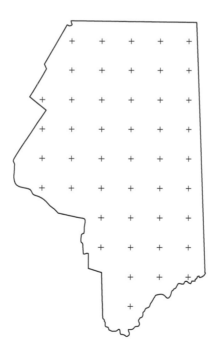

FIGURE 1.3
Lackawanna County locality layer showing 5-km grid of posting points.

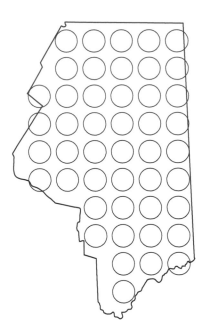

FIGURE 1.4
Provisional proximity perimeters having a 2-km radius for a 5-km pattern of posting points in Lackawanna County.

are informative. Accordingly, Figure 1.5 is a collage (mosaic) of aerial photographs truncated at the boundary. The urban area of Scranton, PA, lies in the elongate Wyoming Valley from middle left to upper right. The Lackawanna River flows through the valley in a southwesterly direction toward a confluence with the Susquehanna River.

For further contextual reference, Figure 1.6 is an excerpt from the digital version (digital raster graphics; DRG) of the U.S. Geological Survey (USGS) 1:100,000 scale quadrangle map covering the Scranton area. It can be seen that the Moosic Mountains lie to the right of Scranton, with Bald Mountain and Bell Mountain to the left. A short segment of the Susquehanna River forming part of the county boundary is included at the lower-left corner. The Pennsylvania Turnpike crosses between Bald Mountain and Bell Mountain.

Figure 1.5 and Figure 1.6 should be bookmarked for later reference in subsequent chapters. For present purposes, compare these two figures to Figure 1.7, which shows the provisional proximity perimeters overlaid on a state road layer (as opposed to a local road layer) for Lackawanna County. The value of Figure 1.5 and Figure 1.6 as interpretive aids should thereby become evident.

FIGURE 1.5
Collage of aerial photographs for Lackawanna County. Note the middle-left to upper-right urbanized area of Scranton, PA.

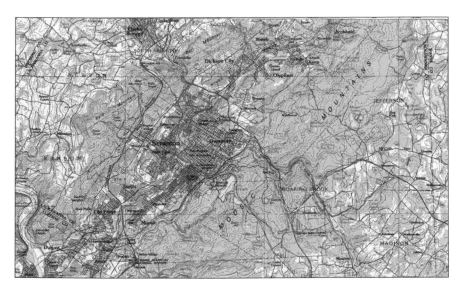

FIGURE 1.6
Excerpt from digital version of USGS 1:100,000 Scranton quadrangle map.

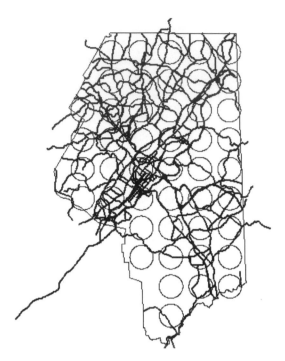

FIGURE 1.7
State roads shown along with provisional proximity perimeters for Lackawanna County.

1.8 Surveillance with Software Sentinels

Instead of passive point postings, statistical sleuthing for spatial structure can be cast in the context of spatial surveillance by agent-like software sentinels, which may help to provide methodological motivation even in the absence of full agent-based formalities. Each observational instance of a posting point is seen as an observatory station for a sentinel system that monitors the status of that situational setting and sends signals via indicators. The structural star in Figure 1.8 conveys this more creative casting of context.

The upward pointing triangular component of the star in Figure 1.8 represents a single soft sentinel system, whereas the downward pointing triangular component represents the distributed discrete domain as a collective context or "society" of sentinels.

Starting at the lower-left of Figure 1.8, we have the solitary sentinel setting as a spatial situation. Moving across to the lower-right, we couple the setting with the signaling in terms of the proximal properties and their scalings. Any data domain is a duality between imposition of instances and selection of signals. The sentinel setting can only be characterized in terms of its

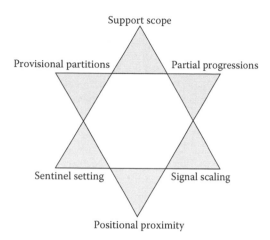

FIGURE 1.8
Structural star of posting points with proximate properties as soft(ware) sentinel surveillance systems.

situational signaling, which is implemented here as indicators. Moving to the apex points out the subtext due to scope of spatial support over which an indicator is integrating information. There is a general propensity for increasing scope of spatial support (enlarged proximities) to encompass greater diversity of detail and so to smooth (average out) the spatial process of an integrative indicator.

Considering the lower corner of the collective context, we have the pervasive potential of spatial auto-association (similarity of nearer neighbors) to influence inference (Bivand et al. 2008). Auto-association typically tends to strengthen with increasing positional proximity and is a central concern of spatial statistics. The differing degrees of such dependence among instances must be incorporated in analytical approaches to seeking spatial structure and purveying propensities. Moving to the upper-right corner, it is of initial interest to investigate progressions or partial progressions among instances of indicators and where there is synchrony in that regard for subsets of signals (indicators). Technical tools obtained from the partial order theory (Bruggemann and Patil 2011) are appropriate here. Interplays among instances and indicators become especially interesting and may produce perspectives whereby specific subsets of instances and indicators portray particular progressions. Even in the absence of prominent progressions, movement over to the upper left may produce provisional partitions among the sentinels whereby there are some collective consistencies within and between the partitions. We might consider this latter as the virtual version of social networking across the distributed domain of sentinels.

1.9 Backdrop: Distributed Data Depots and Digital Delivery

Our functional framework is built on a foundation of digital data that is spatially specific (georeferenced) and often operationally obtainable from designated depositories. These spatial data clearinghouses may be solitary systems or synergistic sets of systems functioning as "active archives" doing distributed delivery. Institutional initiatives often have state subsidy or sponsorship, with the Pennsylvania Spatial Data Access (PASDA) portal for public purposes at www.pasda.psu.edu as a well-established example that is easily explored.

At the national level in the United States, the National Spatial Data Infrastructure (NSDI) provides a formal framework and national network of nodes. The USGS serves as the "front door" facility for NSDI through the National Map (nationalmap.gov) and the Geospatial One Stop (GOS; geodata.gov) portal and seamless server. Spatial data are typically thought of as comprising layers, whereby the content classes of a layer are mostly mutually exclusive in space (except that occurrences of incidents are quite commonly colocated as point placements). NSDI features framework layers that are crucial to human environment interface, such as hydrography, elevation, and land cover along with high-resolution "ortho" (map-matching) imagery.

The U.S. Census Bureau is central to spatially specific information on the human environment interface that meets Federal Information Processing Standards (FIPS). Roads are recorded in their Topologically Integrated Geographic Encoding and Referencing (TIGER) data. Metropolitan and micropolitan areas are concerned with counties, cities, and other minor civil divisions. Census tracts and blocks are used to compile the standard census information.

NASA's Earth Observation Systems (EOS) programs have generated massive amounts of image data at varying levels of spatial resolution, much of which is made available through the NASA Data Centers via the Land Processes Distributed Active Archive Center (LPDAAC; LPDAAC@eos.nasa .gov and nasadaacs.eos.nasa.gov).

Metadata (data about data) are essential in accessing these several types of spatial data clearinghouses. Searches of available information are done in terms of metadata rather than the actual spatial data so as to circumvent difficulties with the disparities in data mentioned earlier. The metadata can be augmented with "thumbnail" portrayals to assist interactive inquiry.

Projections (Iliffe and Lott 2010; Van Sickle 2010) are a problematic part of spatial synthesis to be met in the metadata. Georeferencing requires a coordinate system for positioning and plotting. The coordinate system must conform to the spheroidal surface of the earth and be projected onto a plane for plotting. The earth has the form of an irregularly oblate spheroid which must be given a first approximation by a mathematical model—of which there are several having different preferential properties but lacking an overall optimum. The spheroid or ellipsoid model gives rise to a "geoid" as a gravitational equipotential surface that defines nominal sea level. Remaining irregularities are accommodated somewhat empirically in a datum that models local variations in the spheroid/geoid. A common coordinate system among several layers of data sets the stage for spatial synthesis. The process for procurement thus entails "connect and project" then "clip and ship," whereby compatible coordinates are clipped to conform to the area of interest. The sophistication of the server system for a clearinghouse is reflected in its ability or lack thereof for projecting and clipping. If these operations are not automated on the server side, then they will need to be done on a postprocurement basis. Data recorded in terms of latitude and longitude are said to lack projection because convergence of longitude at different levels of latitude will distort planar plotting. A further advantage of point purviews for soft sentinels is that perspectives of one projection can be transferred to another for comparative purposes.

References

Bivand, R., Pebesma, E., and Gomez-Rubio, V. *Applied Spatial Data Analysis with R.* New York: Springer, 2008.

Bruggemann, R., and Patil, G. P. *Ranking and Prioritization for Multi-indicator Systems.* New York: Springer, 2011.

Campagna, M., Ed. *GIS for Sustainable Development.* Boca Raton, FL: Taylor & Francis/ CRC, 2006.

DeMers, M. *Fundamentals of Geographic Information Systems.* New York: Wiley, 2009.

Ehlers, M. Geoinformatics and digital earth initiatives: a German perspective. *International Journal of Digital Earth* **1**(1): 17–30, 2008.

Forman, R. T. *Urban Regions: Ecology and Planning Beyond the City.* New York: Cambridge Univ. Press, 2008.

Fortin, M. J., and Dale, M. *Spatial Analysis: A Guide for Ecologists.* Cambridge, UK: Cambridge Univ. Press, 2005.

Iliffe, J., and Lott, R. *Datums and Map Projections for Remote Sensing, GIS and Surveying, 2nd ed.* Dunbeath, Scotland: Whittles Publishing, 2010.

Joza, A., and Brown, D. Neighborhood Sustainability Indicators Report on a Best Practice Workshop. School of Urban Planning, McGill University and the Urban Ecology Center/SodecM, Montreal, June 10–11, 2005.

Khavari, F. A. *Environomics: The Economics of Environmentally Safe Prosperity.* Westport, CT: Praeger, 1993.

Luloff, A., Finley, J., Myers, W., Metcalf, A., Matarrita, D., Gordon, J., Raboanarielina, C., and Gruver, J. What do stakeholders add to identification of conservation lands? *Society and Natural Resources* **24**: 1345–1353, 2011.

Maguire, D., Goodchild, M., and Batty, M., Eds. *GIS, Spatial Analysis, and Modeling.* Redlands, CA: ESRI Press, 2005.

McMaster, R., and Usery, E. L. *A Research Agenda for Geographic Information Science.* Boca Raton, FL: Taylor & Francis, 2004.

Myers, W., and Patil, G. P. Biodiversity in the age of ecological indicators. *Acta Biotheoretica* **54**: 119–123, 2006.

Obermeyer, N., and Pinto, J. *Managing Geographic Information Systems, 2nd ed.* New York: Guilford, 2008.

Panigrahi, N. *Geographical Information Science.* Boca Raton, FL: Taylor & Francis/CRC, 2010.

Van Sickle, J. *Basic GIS Coordinates, 2nd ed.* Boca Raton, FL: CRC Press, 2010.

2

Localizing Fixed-Form Features

2.1 Introduction

Contexts of concern will often involve spatial "layers" of formed features consisting of coordinates or connected coordinates with associated attributes. The coordinates will consist of coupled latitude and longitude values or pairs on planar projections. Each feature will have an individual identifier indexing a line of a spreadsheet with the associated attributes as columns. A feature's form will be that of a point, a path, or patch(es). These three types will be segregated in layers, with a layer consisting entirely of points, entirely of paths, or entirely of patches. The types are considered as "topologies" in the terminology of geographic information systems (GIS). These will be "thematic" layers, with the features in a given layer pertaining to some subject. Patches in a particular layer will be spatially adjacent or isolated but not overlapping. In a typical GIS, the topological information regarding spatial structure is stored separately from the attribute information, with a feature identifier (FID) serving as linkage (see Section 2.7). In this chapter, we pursue particulars of our localization paradigm as applied to fixed-form features using Lackawanna County as a laboratory.

2.2 Locality Layer as Poly-Place Purview

The point of departure for our surveillance strategy is to generate a square pattern of points as a *locality layer* spanning an extent of interest. A *vicinity variate* as an item of interest is chosen to receive a value obtained in proximity of each point (locality) and attributed to the point as a proximate property. A collective of such points and their proximate properties comprises a *poly-place purview*. Such a poly-place purview constitutes the construct of spatial structure that is most conformant with conventional observational instances wherein each instance comprises a row of a data table (data frame) with properties (variables) as columns. Even though point coordinates for localities may be kept in a separate database by a GIS, the couplet of coordinates can also augment the table as additional columns so that the table is self-standing as integrated information. It is, of course, necessary to avoid applying conventional multivariate methods to the coordinate columns. Thus, we begin with a pattern of points to serve as a locality layer as already introduced for Lackawanna County in the first chapter.

There are several strategies for generating coordinates of points for a locality layer. Perhaps the simplest of such strategies is to begin with the corner coordinates of a bounding box for the area of interest. A grid of points can then be generated by adding appropriate increments from the corners in a suitable software facility. The collection of coordinates can then be output in a generic file form such as delimited text for transfer to other software systems as needed. An alternative is to start with a boundary for an irregular area of interest and generate points within the boundary.

Commercial ArcGIS© software by Environmental Systems Research Institute (ESRI, Redlands, CA) (Kennedy 2009; Ormsby et al. 2010) provides the primary platform for the Lackawanna example, working from county boundary data downloaded from the Pennsylvania Spatial Data Access (PASDA) clearinghouse as introduced in the previous chapter. ArcGIS has a "fishnet" facility for generating a grid of lines and placing points at the centers of the grid squares as anchors for labels. The points are separable and can serve our purpose. Points external to a boundary can then be clipped away as a subsequent step. However, the set of points shown earlier in Figure 1.3 and considered here was generated in a different facility called fGIS (see www.forestpal.com website) as survey (cruise) points and subsequently used with ArcGIS. This (poly-point) set of localities is shown in Table 2.1, with coordinate units being meters.

The attribute table shown in a GIS facility may contain some columns that do not transfer readily by "export" to outside facilities, and the coordinate columns may be missing from the table due to storage of topologies in binary database domains. For example, the first column of Table 2.1 becomes problematic for external spreadsheet systems like Excel©. Notice that the numbering in the FID column starts at 0 whereas the numbering in the ID column starts at 1. The FID is an internal index for ArcGIS to link point features with their attributes. The FID component appears in exports only as a column of commas, which confounds the ID column in Excel. Likewise, there is a column in the ArcGIS table that declares each row as a POINT, which does not appear at all in exports. It is necessary to invoke a special request in ArcGIS to have the coordinates included in the table as shown here. We mention such details here because they can be quite frustrating if not anticipated, even for those who are moderately acquainted with a GIS facility. The ID column numbered from 1 is particularly critical to our overall strategy because it is the means for matching several sets of point properties to create a multivariate poly-place purview that is transportable to statistical software.

TABLE 2.1

Locality Point Placements at 5-km Spacing for Lackawanna County

FID	ID	POINT_X	POINT_Y
0	1	450265.0626	4561637.415
1	2	450265.0626	4566637.415
2	3	455265.0626	4566637.415
3	4	460265.0626	4566637.415
4	5	445265.0626	4571637.415
5	6	450265.0626	4571637.415
6	7	455265.0626	4571637.415
7	8	460265.0626	4571637.415
8	9	445265.0626	4576637.415
9	10	450265.0626	4576637.415
10	11	455265.0626	4576637.415
11	12	460265.0626	4576637.415
12	13	435265.0626	4581637.415
13	14	440265.0626	4581637.415
14	15	445265.0626	4581637.415
15	16	450265.0626	4581637.415
16	17	455265.0626	4581637.415
17	18	460265.0626	4581637.415
18	19	435265.0626	4586637.415
19	20	440265.0626	4586637.415
20	21	445265.0626	4586637.415
21	22	450265.0626	4586637.415
22	23	455265.0626	4586637.415
23	24	460265.0626	4586637.415
24	25	435265.0626	4591637.415
25	26	440265.0626	4591637.415
26	27	445265.0626	4591637.415
27	28	450265.0626	4591637.415
28	29	455265.0626	4591637.415
29	30	460265.0626	4591637.415
30	31	435265.0626	4596637.415
31	32	440265.0626	4596637.415
32	33	445265.0626	4596637.415
33	34	450265.0626	4596637.415
34	35	455265.0626	4596637.415
35	36	460265.0626	4596637.415
36	37	440265.0626	4601637.415
37	38	445265.0626	4601637.415
38	39	450265.0626	4601637.415

(continued)

TABLE 2.1 (Continued)

Locality Point Placements at 5-km Spacing for Lackawanna County

FID	ID	POINT_X	POINT_Y
39	40	455265.0626	4601637.415
40	41	460265.0626	4601637.415
41	42	440265.0626	4606637.415
42	43	445265.0626	4606637.415
43	44	450265.0626	4606637.415
44	45	455265.0626	4606637.415
45	46	460265.0626	4606637.415

Figure 2.1 shows the locality layer with points labeled by their ID numbers. If points are generated beyond the boundaries and later clipped away, the remaining points should be renumbered in sequence by using rank numbers obtained from remaining IDs.

In more sophisticated scenarios, one might generate a double-density locality layer and then separate it into two half-density subsets with dual numberings. One numbering would be sequential across subsets, and the other numbering would be sequential within subsets. The two numberings would appear in different columns of the data table.

FIGURE 2.1
Lackawanna County locality layer with ID numbers for posting points.

2.3 Localizing Layer of Proximity Perimeters

Zones of proximity around points as well as other formal features can be based on buffering operations in GIS. A buffer around a point approximates a circle by connected chords forming a polygonal feature. The series of connected chords closes on its point of beginning, as is the case for any polygonal patch. Buffering operations are available in most GIS software systems. Figure 2.2 shows circular buffers having 2-km radii around the posting points in Lackawanna County. The circular buffers shown in Figure 2.2 were generated with ArcGIS, but the figure was rendered with MapWindow© (which is an open software system) to illustrate a variety of software facilities.

At this juncture and with a seemingly small set of 46 proximities, we can begin to appreciate the sophistication of commercially capable GIS software systems. The chord connections for each of the 46 circular buffers entail 315 pairs of point coordinates, and repetition of the beginning makes 316. Table 2.2 contains a textual listing for the first six and last six points in one of the buffers. With only 46 circular buffers but each having 316 points, there are 14,536 pairs of point coordinates.

It thus becomes important to consider the implications of different modes of storing coordinates, buffer geometries, and/or degrees of thinning for coarser chords. An American Standard Code for Information Interchange (ASCII) (textual) listing as in Table 2.2 requires one byte of computer storage

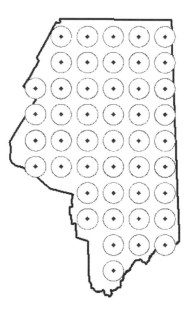

FIGURE 2.2
Circular buffers with 2-km radii around points in Lackawanna County.

TABLE 2.2

Textual Listing Showing Leading and
Trailing Six Points for One Circular Buffer

Polygon		
0 0		
0	450265.062552	4563637.41479
1	450285.072362	4563637.31469
2	450325.083972	4563636.51395
3	450365.071549	4563634.91278
4	450405.019083	4563632.51183
5	450444.910579	4563629.31205
6	450484.730066	4563625.31474
.		
.		
.		
310	450085.214525	4563629.31205
311	450125.106021	4563632.51183
312	450165.053555	4563634.91278
313	450205.041132	4563636.51395
314	450245.052742	4563637.31469
315	450265.062552	4563637.41479

capacity for each character. It is far more parsimonious to store the coordinates in a binary mode so that only a couple of bytes are needed for each coordinate. The latter situation is a major reason that topologies are stored in a different database domain than the attributes, which require only one line in a table for each feature regardless of the topological type.

Consider also an alternative of having octagonal proximity perimeters, which would then entail only $46 \times 9 = 414$ pairs of coordinates. An inscribed octagon encompasses 89.898% of the area of the circle because its area is $2.82424 \times R^2$, where R is the radius of the circle. It is also a straightforward computational matter to generate an octagon around a point because the sizes of the increments from the center point to the corners are either R or $0.707107 \times R$. However, most GIS facilities do not provide for generating octagonal zones around points. Thus, octagonal geometry presents the prospect of creating a special facility for generating the proximity perimeters. Therefore, we defer the consideration of octagonal integration vicinities (OCTIVs) until later.

Fortunately, proximity perimeters are transient topologies in localizing scenarios. Once the indicator(s) serving as proximate properties are composed, they are transferred as attributes to the posting points. Thereafter, the layer(s) of proximity perimeters can be relegated to off-line storage if it is deemed desirable to retain them.

It was noted in the previous chapter that circular proximity perimeters are provisional because some of them straddle county boundaries. A boundary bias will be incurred when integrated vicinity indicators (IVIs) are compiled over a fixed-radius vicinity that extends beyond the boundary of an area of interest.

There are several more or less minor GIS facilities for mapping, but more major facilities also perform geometric/geographic analysis involving interplay of spatial topologies from different layers. A major genre of geometric analyses entails various overlay operations, one of which is called the "clipping" (or truncating) of features in one layer using features from a polygonal layer. Clipping is to be distinguished from "intersecting" since clipping simply truncates formed features, whereas intersecting is also combinatorial in terms of attributes. Figure 2.3 shows circular buffers of Figure 2.2 clipped at the county boundary so that external portions are eliminated. These are now valid vicinities that comprise a *localizing layer*.

Areas of the circular buffers were not shown explicitly in the attribute table before clipping but were easily calculated as the constant area for 2-km-radius circles. The areas of clipped components, however, are not so conveniently constant. In the particular version of ArcGIS used here, this is accomplished with field calculator facilities in the attribute table side of the GIS. The opened attribute table includes an unobtrusive "Options" button giving access to "Table Tools," whereby a new (empty) column (field) can be

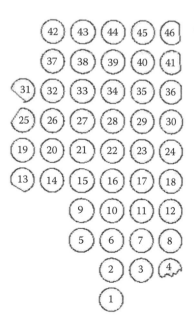

FIGURE 2.3
Clipped proximity perimeters for Lackawanna County.

added to the table (call it PParea) with suitable type and precision (here double). Highlighting the head of the new column and right-clicking then offers the option to "Calculate Geometry." With "Area" being chosen and suitable units (here square meters), the areas of the buffers (after clipping) are calculated and appear in the column. These details are noted here to show where potential pitfalls may arise unless the services of a well-trained GIS analyst are available. The message is that a team effort is often needed.

2.4　Localizing Linears by Determining Densities

Having obtained proximity perimeters that are suitably constrained to lie within the area of interest (here Lackawanna County), we can undertake to compile indicators of interest as vicinity variates. Let us pursue the complex concept of accessibility inasmuch as this is of interest both for habitation and for ecology/environment. Regional (state) roads will serve as our exemplar, with local roads being handled in a parallel manner. Figure 1.7 provides visual impetus for constructing an IVI. Road densities (length per unit area) offer one possible path of pursuit.

Toward this end, we need to *isolate* the respective roads within proximity perimeters and determine the lengths of these portions. The first step of isolation is another application of clipping, whereby the linear layer of roads is clipped to the proximity perimeters. The isolated regional roads are shown in Figure 2.4, and the isolated local roads are shown in Figure 2.5.

With ArcGIS, a "Length" column (field) will be in the attribute table but will not have actual feature lengths computed. To accomplish the computation of feature lengths in this situation, one must utilize the tabular "field" calculator facilities by clicking on the head of the column to highlight it and then right-clicking to get the calculator menu and choosing "Calculate Geometry."

Clipping linear features by proximity perimeters does not associate a particular road with a particular proximity because external portions are simply purged. Associating remaining roads with proximities is accomplished by an intersection operation whereby each clipped road has an attribute identifying the proximity in which it occurs. At this stage, it is important to note that the features are of the linear type, with each being a portion of a road that lies within a particular proximity.

IVIs for the road densities can be formulated as total length of road type for the vicinity divided by area of the vicinity. Length and area should have corresponding scale units, which are meters and square meters in the current case. Unit cancellation in numerator and denominator will give units for the indicator as inverse meters. Because these numbers have small magnitudes,

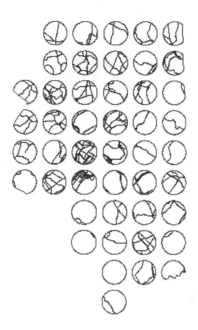

FIGURE 2.4
Isolation (by clipping) of regional (state) roads within proximity perimeters for Lackawanna County.

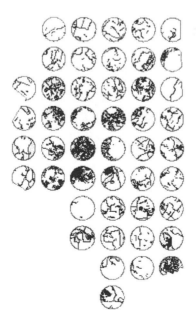

FIGURE 2.5
Isolation (clipping) of local roads to proximity perimeters in Lackawanna County.

it will be desirable to rescale the indicators to inverse kilometers by multiplying them by 1000.

Total lengths can be obtained using the table tools of ArcGIS by opening the attribute table, clicking on the name of the proximity patch field to highlight it, right-clicking to get the field menu, choosing "Summarize," and then designating length for summarizing as a sum statistic. This creates a (new) summary table that can be joined to the parent table.

Since the division has not yet been performed, it will be convenient to transfer the areas of the proximity patches to make this new table self-contained with regard to the indicator information. This requires adding a new column to the table having the same type as the area column of the parent table and then copying the information into the column from the parent table after joining the parent table.

A new indicator column "RrdsIVI" or "LrdsIVI" can then be added to each of the respective tables and calculated with the field calculator as the quotient of the sum column and the area column. The field (column) calculator is then used again to rescale the indicators by multiplying themselves by 1000.

2.5 Transfer from Perimeters to Points

Inasmuch as the identifiers of the posting points correspond exactly to the identifiers of the proximity perimeters, it is a simple matter to make a transient juxtaposition of road IVI on posting point by joining the attribute table containing IVI to the attribute table for posting points in the GIS. This is shown for local roads in Figure 2.6 using graduated symbols to represent road density.

To make a permanent transfer of the IVI data to posting points, new columns (fields) are added to the attribute table for posting points. Appropriate columns for this situation are PParea for area in proximity perimeter, RrdIVI for the regional (state) road IVI, and LrdIVI for the local road IVI. These new columns must match the corresponding column types for the attribute table(s) in which the desired data currently reside. In separate operations, each of the source tables is then *joined* to the posting point table and the field calculator facilities used to copy values from the respective column in the source table to the skeleton column in the posting point table. The respective table join is removed when the copy operation has been completed. Figure 2.7 shows the IVI for regional (state) road density from the final table using proportional symbols. Note that one locality in the lower end of the county has no such regional roads.

The attribute table for posting points is now ready to be exported from the GIS as a generic textual file. For the current context, exporting takes place

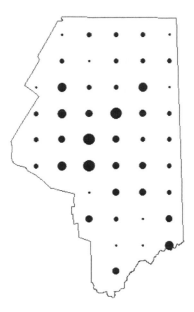

FIGURE 2.6
IVI for local road density shown by graduated symbols for Lackawanna County.

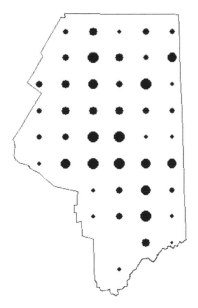

FIGURE 2.7
IVI for regional (state) road density shown by proportional symbols for Lackawanna County.
Note absence of such roads for posting point number 2.

TABLE 2.3

Leading Lines of CSV Table for Road Density Indicators

```
"FID_","ID","POINT_X","POINT_Y","PParea","RrdIVI","LrdIVI"
,"1",450265.062552,4561637.414790,12565534.026000,0.298855,2.366882
,"2",450265.062552,4566637.414790,12565534.026000,0.000000,0.747736
,"3",455265.062552,4566637.414790,12565534.026000,1.956432,0.688363
,"4",460265.062552,4566637.414790,8754482.111960,0.104275,5.107497
```

in "comma separated values" (CSV) format, with Table 2.3 showing the first few lines.

There are several things to be noted in Table 2.3 so that they can be treated appropriately when imported into a flexible external spreadsheet facility such as Excel. The first line contains the names of the columns (fields), and there appears to be one more name than fields on the following lines. The first name as "FID_" is the internal feature ID and is represented in subsequent lines only by a comma. In the parent table, this column started at zero and increased in unit steps. The second field is ID, but the values are enclosed in quotes on subsequent lines, implying that these are treated in the manner of names rather than numbers. When being imported into Excel, the "FID_" can be an empty column to receive serially increasing numbers by direct entry. Likewise, the ID field can be designated for numeric conversion if desired. This is displayed in Table 2.4 after supplying the entries for the "FID_" initial column. After being configured in Excel, the data can be re-exported as a more conventional text file (leading lines shown in Table 2.5) for importing into a statistical software system. Initial analysis of these data is undertaken in the next chapter.

TABLE 2.4

Excel Table Obtained by Importing External Data from File of Table 2.3

FID_	ID	POINT_X	POINT_Y	PParea	RrdIVI	LrdIVI
0	1	450265.0626	4561637.415	12565534.03	0.298855	2.366882
1	2	450265.0626	4566637.415	12565534.03	0	0.747736
2	3	455265.0626	4566637.415	12565534.03	1.956432	0.688363
3	4	460265.0626	4566637.415	8754482.112	0.104275	5.107497
4	5	445265.0626	4571637.415	12565534.03	0.319571	2.208935
5	6	450265.0626	4571637.415	12565534.03	1.110692	1.291283

(continued)

TABLE 2.4 (Continued)

Excel Table Obtained by Importing External Data from File of Table 2.3

FID_	ID	POINT_X	POINT_Y	PParea	RrdIVI	LrdIVI
6	7	455265.0626	4571637.415	12565534.03	3.429123	0.963902
7	8	460265.0626	4571637.415	12565534.03	0.332491	2.771196
8	9	445265.0626	4576637.415	12565534.03	0.319571	0.192168
9	10	450265.0626	4576637.415	12565534.03	1.113059	1.926716
10	11	455265.0626	4576637.415	12565534.03	2.485752	2.29541
11	12	460265.0626	4576637.415	12565534.03	0.863636	1.585662
12	13	435265.0626	4581637.415	11765485.43	0.443674	1.271092
13	14	440265.0626	4581637.415	12565534.03	2.435064	4.73941
14	15	445265.0626	4581637.415	12565534.03	3.120936	7.17082
15	16	450265.0626	4581637.415	12565534.03	3.529457	2.786172
16	17	455265.0626	4581637.415	12565534.03	2.992038	1.930051
17	18	460265.0626	4581637.415	12565534.03	2.666149	1.280703
18	19	435265.0626	4586637.415	12565534.03	0.955322	1.178856
19	20	440265.0626	4586637.415	12565534.03	1.484586	2.84826
20	21	445265.0626	4586637.415	12565534.03	3.426207	8.726684
21	22	450265.0626	4586637.415	12565534.03	3.822582	2.270837
22	23	455265.0626	4586637.415	12565534.03	0.475379	1.67072
23	24	460265.0626	4586637.415	12565534.03	0.477585	1.766848
24	25	435265.0626	4591637.415	11535256.65	0.993364	1.123384
25	26	440265.0626	4591637.415	12565534.03	2.055861	3.627327
26	27	445265.0626	4591637.415	12565534.03	2.253253	2.159982
27	28	450265.0626	4591637.415	12565534.03	2.224338	6.743757
28	29	455265.0626	4591637.415	12565534.03	1.566584	2.340805
29	30	460265.0626	4591637.415	12518880.39	0.843508	1.594994
30	31	435265.0626	4596637.415	10477994.69	1.06897	0.820908
31	32	440265.0626	4596637.415	12565534.03	2.195724	3.495175
32	33	445265.0626	4596637.415	12565534.03	2.996226	1.65802
33	34	450265.0626	4596637.415	12565534.03	1.203704	1.665202
34	35	455265.0626	4596637.415	12565534.03	3.085494	3.924797
35	36	460265.0626	4596637.415	12177219.58	0.336239	0.440188
36	37	440265.0626	4601637.415	12565534.03	1.135701	1.480685
37	38	445265.0626	4601637.415	12565534.03	3.176763	1.036531
38	39	450265.0626	4601637.415	12565534.03	2.116433	1.179013
39	40	455265.0626	4601637.415	12565534.03	0.941875	1.814449
40	41	460265.0626	4601637.415	11662930.51	2.701422	1.772773
41	42	440265.0626	4606637.415	12565534.03	0.895236	0.647019
42	43	445265.0626	4606637.415	12565534.03	2.162687	1.113021
43	44	450265.0626	4606637.415	12565534.03	0.614937	1.414668
44	45	455265.0626	4606637.415	12565534.03	1.118016	1.317684
45	46	460265.0626	4606637.415	11278387.96	0.872076	0.606927

TABLE 2.5

Leading Lines of File Saved from Excel as Tab-Delimited Text with Name of First Column Edited to Remove Trailing Underscore

FID	ID	POINT_X	POINT_Y	PParea	RrdIVI	LrdIVI
0	1	450265.0626	4561637.415	12565534.03	0.298855	2.366882
1	2	450265.0626	4566637.415	12565534.03	0.000000	0.747736
2	3	455265.0626	4566637.415	12565534.03	1.956432	0.688363
3	4	460265.0626	4566637.415	8754482.112	0.104275	5.107497
4	5	445265.0626	4571637.415	12565534.03	0.319571	2.208935
5	6	450265.0626	4571637.415	12565534.03	1.110692	1.291283
6	7	455265.0626	4571637.415	12565534.03	3.429123	0.963902

2.6 Apportioning Attributes of Partial Polygons

Although our effort in this chapter has been directed toward localizing linear features, a parallel process applies to polygonal features. A crucial caveat in this regard is that most attributes of polygons need to be apportioned by area when polygons are split by clipping or intersection during localization. The key to avoiding difficulty is to cast attributes to be apportioned as densities and rates *before* embarking on localization. For example, population in a polygon would be cast as density per unit area instead of quantity present.

2.7 Backdrop: GIS Generics

As described at the beginning of the chapter and continuing throughout, fixed-form features have dual aspects of locational topology and attribute characteristics. The former consists of coordinates, and the latter appear as tabular columns (Bernhardsen 2002; Burrough and McDonnell 1998; Chrisman 2002; Demers 2009; Harvey 2008; Longley et al. 2005). A GIS must link these two aspects dynamically, but there tends to be some separation of topological tools and tabular tools. There is mutual advantage in having at least one format for spatial data that accommodates different topologies and can be a de facto standard for information exchange. So-called shape files that originated with the predecessors of ArcGIS have effectively emerged as an informal standard in this respect. By "generic GIS" we mean those that are geared to handle fixed-form features from shape files as a major mode of operation. However, this will not necessarily be their only mode of managing spatial data. For instance, even though shape files originated with ESRI, their ArcGIS is moving more toward a "geodatabase" mode, and prior to shape files, they used a "coverage" mode involving a tighter topology based more on networks of features than individuality of features.

It is important to know that shape files are not monolithic but rather a suite of files that share a common root name with different extensions and functionality. One of these files having a ".shp" extension actually holds the binary topological information on the shape or form of the feature. A file with a ".dbf" extension as a legacy form of DBASE holds the tabular attribute information. A file with a ".shx" extension serves for cross-indexing topology and attributes. A file with a ".prj" extension holds the geographic projection information for mapping. These different files must be in the same folder (directory) but are otherwise relatively transportable by copying from one folder to another. However, some other auxiliary files often present are not necessarily so transportable.

Along with shape files has come a legacy of layout for the desktop that a user sees for interacting in a map-like mode with geospatial data. For this illustrative introduction, we look to the open-source MapWindow system (www.mapwindow.org) as a model. Figure 2.8 shows the MapWindow desktop for our Lackawanna County setting.

The desktop in Figure 2.8 has a window on the right-hand side where the features are displayed as a map. On the left-hand side is a "Legend" window that serves as an index to the layers of features that are active. At the top is a line of drop-down menus such as "File" and "Edit." A menu is made to appear by clicking on its name. Below the line of menus is a set of tools that

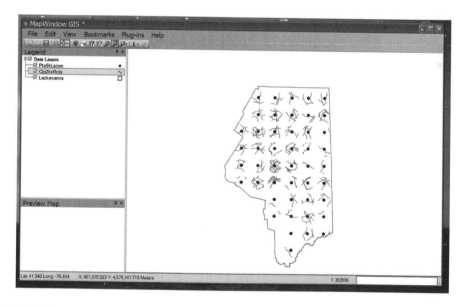

FIGURE 2.8
MapWindow desktop showing locality points, county boundary, and regional (state) roads within 2-km-radius proximities.

are operative when the cursor is in the mapping window. These tools typically include a "cross" icon for adding new map layers, "looking-glass" icons for zooming the map in and out, a "hand" icon for panning by pulling the map, an "i" icon for getting information about a feature, and so on. A tool is activated by clicking on its icon and then moving the cursor over the map. MapWindow has a "Preview Map" window at the lower-left hand, which will not necessarily be found in that place of the desktop for other GIS systems and can be closed with the corner "x" of this desktop. Most such GIS systems work in terms of "projects," with a project being the current configuration of the desktop so that work can be saved and continued from one session to the next. Menu and button options are available for opening the companion attribute table in a window, after which the table tools are much like those of a spreadsheet. Access to topological tools varies considerably between systems and even from one version of a system to another. Systems often bundle their topological tools as add-ons that must be acquired and installed separately. Some of such add-ons are contributed by outside developers.

MapWindow subscribes to an open software philosophy and is joined in this sentiment by the Open Geospatial Consortium (OGC; www.opengeospatial .org). Several other geospatial developers also have some component of their software that is available without cost. There is a companion freeware viewer to ArcGIS that is called ArcGIS Explorer (for further information, do an Internet search on "ArcGIS Explorer"). The fGIS facility (www.forestpal.com) has been mentioned earlier. The Geographic Resources Analysis Support System (GRASS) GIS is publicly available software of a somewhat different nature.

References

Bernhardsen, T. *Geographic Information Systems: An Introduction*. New York: Wiley, 2002.

Burrough, P., and McDonnell, R. *Principles of Geographical Information Systems*. New York: Oxford Univ. Press, 1998.

Chrisman, N. *Exploring Geographic Information Systems*. New York: Wiley, 2002.

Demers, M. *Fundamentals of Geographic Information Systems*. New York: Wiley, 2009.

Harvey, F. *A Primer of GIS: Fundamental Geographic and Cartographic Concepts*. London: Guilford Press, 2008.

Kennedy, M. *Introducing Geographic Information Systems with ArcGIS: A Workbook Approach to Learning GIS, 2nd ed*. New York: Wiley, 2009.

Longley, P., Goodchild, M., Maguire, D., and Rhind, D. *Geographic Information Systems and Science, 2nd ed*. New York: Wiley, 2005.

Ormsby, T., Napolean, E., Burke, R., Groessl, C., and Bowden, L. *Getting to Know ArcGIS Desktop*. Redlands, CA: ESRI Press, 2010.

3

Precedence and Patterns of Propensity

3.1 Introduction

Having developed two vicinity variates as integrated vicinity indicators (IVIs) around posting point localities to address the somewhat abstract notion of accessibility, we proceed to investigate patterns of propensity in this regard through precedence and progression (Myers and Patil 2010, 2012). Inasmuch as there is more than one indicator involved, all indicators must be honored in this respect. The approach that is to be presented can accommodate multiple indicators as needed, and it is not disturbed by spatial auto-association in or among indicators. We rely only on ordinal information in the indicators and work empirically without attempting to address significance or lack thereof at this stage. We transfer the localized data into the **R**© statistical software system (see Section 3.9) from a text file for conducting this analysis. In so doing, we give the relevant **R** commands and outputs so that the approach is completely specified. In this chapter, we do not use any special packages for **R** aside from the function facilities presented here. Accordingly, the first concern is to import information as a data frame object into **R**.

```
> LacwnRds <- read.table("Lacwn5n2rds.txt",header=T)
> names(LacwnRds)
[1] "FID"     "ID"      "POINT_X" "POINT_Y" "PParea"  "RrdIVI"  "LrdIVI"
> head(LacwnRds)
  FID ID POINT_X POINT_Y   PParea    RrdIVI   LrdIVI
1   0  1 450265.1 4561637 12565534 0.298855 2.366882
2   1  2 450265.1 4566637 12565534 0.000000 0.747736
3   2  3 455265.1 4566637 12565534 1.956432 0.688363
4   3  4 460265.1 4566637  8754482 0.104275 5.107497
5   4  5 445265.1 4571637 12565534 0.319571 2.208935
6   5  6 450265.1 4571637 12565534 1.110692 1.291283
> tail(LacwnRds)
   FID ID POINT_X POINT_Y   PParea    RrdIVI   LrdIVI
41  40 41 460265.1 4601637 11662931 2.701422 1.772773
42  41 42 440265.1 4606637 12565534 0.895236 0.647019
43  42 43 445265.1 4606637 12565534 2.162687 1.113021
44  43 44 450265.1 4606637 12565534 0.614937 1.414668
45  44 45 455265.1 4606637 12565534 1.118016 1.317684
46  45 46 460265.1 4606637 11278388 0.872076 0.606927
```

It is helpful to have scatterplots of the indicators by pairs for visualization, and we exercise the lattice plot facility by including the ID of posting points in the plotting to produce Figure 3.1.

```
> pairs(LacwnRds[,c(2,6,7)])
```

While there is some indication of correlation between the indicators, there are several locations with high regional (state) road density that have relatively low local road density. The Pearson and Spearman correlations are as follows, showing that these two indicators have relatively low redundancy.

```
> cor(LacwnRds$Rrd,LacwnRds$Lrd,method="pearson")
[1] 0.3739562
> cor(LacwnRds$Rrd,LacwnRds$Lrd,method="spearman")
[1] 0.3137315
```

The lower-left corner of Figure 3.1 shows that the localities having highest local road density occur in the middle of the county according to the ID numbering. The specific localities having the higher local road densities can be identified on a simple scatterplot of the two indicators as in Figure 3.2. Localities 15 and 21 are seen to have high accessibilities in both respects.

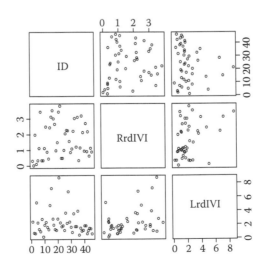

FIGURE 3.1
Lattice of scatterplots for IVIs and their IDs for Lackawanna County.

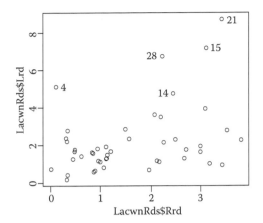

FIGURE 3.2
Scatterplot of IVIs with points having higher local roads identified.

```
> plot(LacwnRds$Rrd,LacwnRds$Lrd)
> identify(LacwnRds$Rrd,LacwnRds$Lrd)
```

For comparative purposes, it is also useful to carry forward the map of ID numbers from Figure 2.2 as shown in Figure 3.3, remembering that the ID numbers start at one whereas the FID numbers start at zero. Numbering

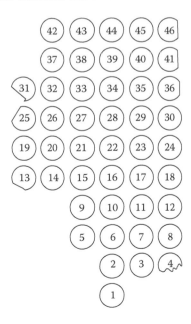

FIGURE 3.3
Clipped proximity perimeters with ID numbers for Lackawanna County.

starts at the southern tip of the county and increases from west to east along the rows of localities. From this, it can be seen that the localities numbered 15 and 21 are adjacent in a north–south line just to the west of center in the county.

3.2 Prescribing Precedence

We seek to allow for flexibility, adaptation, and innovation in our approach to precedence. Therefore, we assume that a protocol will be provided for prescribing precedence or lack thereof between members of a pair of informational instances, which in this situation are localities for IVIs. We provide an **R** function adapted from Myers and Patil (2010) for one such protocol known in partial order theory (Bruggemann and Patil 2010) as *product– order relation* to fill this role in our exposition. At this juncture, we assume availability of such a protocol referencing a pair of instances with multiple indicators and determining that either (1) one member of the pair has precedence over the other, or (2) this pair of instances is indefinite with respect to precedence according to the indicators. For a particular pair, we will thus have

(1) Instance i has pairwise precedence (pp) over instance j with the reciprocal relation being one of definite deficiency (dd);
(2) Instances i and j are pairwise indefinite instances (ii) for which precedence is not assigned by the protocol.

Each of the instances can then be treated as focus of comparison by pairing with each of the other cc = $n–1$ *competing cases* to tabulate the following, where *ff*() denotes the focal frequency of pairwise occurrence.

$$PP = 100 \times ff(\text{pp})/cc$$

$$DD = 100 \times ff(\text{dd})/cc$$

$$II = 100 \times ff(\text{ii})/cc$$

PP is thus the *precedence purview* and DD + II is the *deficiency domain*, with DD being the definite part of that domain.

3.3 Product–Order Precedence Protocol

Precedence is prescribed by the product–order protocol for instance *i* over instance *j* if *i* has no indicators superseded and at least one indicator superseding relative to *j*. The sense of superseding depends on the polarity of the indicators. In other words, *i* must be better in at least one respect and cannot be worse in any respect. An **R** function to tabulate precedence according to the product–order protocol is given as Function 3.1.

Function 3.1 Product–order precedence function for R.

```
POprecdn <- function(CaseIDs,Ratings,BigrBetrl)
 {Ncol <- length(Ratings)
  # CaseIds is vector of IDs for cases.
  # Ratings is a data frame of ratings.
  # BigrBetrl = 1 means bigger is better.
  # First column of output is case ID.
  # Second column of output is percent precedence.
  # Third column of output is percent definite deficiency.
  Ncase <- length(Ratings[,1])
  Placing <- 1
  if(BigrBetrl==1) Placing <- 0
  Status1 <- rep(-1,Ncase)
  Status2 <- Status1
  Prs <- Ncase - 1
  for(I in 1:Ncase)
   {Nosub <- 0; Levl <- 0
     for(J in 1:Ncase)
      {if(I<J | I>J)
        {MatchA <- 0; MatchB <- 0; Undom <- 1
          VecA <- Ratings[I,] - Ratings[J,]
          if(Placing>0) VecA <- Ratings[J,] - Ratings[I,]
          if(max(VecA) > 0) MatchA <- 1
          if(min(VecA) < 0) MatchB <- 1
          if(MatchA==1 & MatchB==0) Nosub <- Nosub + 1
          if(MatchA==0 & MatchB==1) Undom <- 0
          Levl <- Levl + Undom
         }
      }
    Status1[I] <- Nosub
    Status2[I] <- Prs - Levl
   }
  PP <- Status1
  DD <- Status2
  Pct <- 100/Prs
  PP <- PP * Pct
  PP <- round(PP,digits=2)
  DD <- DD * Pct
  DD <- round(DD,digits=2)
  Precidnc <- cbind(CaseIDs,PP,DD)
  return(Precidnc)
 }
```

The POprecdn function is applied as follows:

```
> BigBetr <- 1
> PlacIDs <- LacwnRds[,2]
> RdsPrecdn <- POprecdn(PlacIDs,LacwnRds[,6:7],BigBetr)
```

We can then check the results for localities 15 and 21 noted earlier.

```
> RdsPrecdn[c(15,21),]
     CaseIDs    PP    DD
[1,]      15 88.89  2.22
[2,]      21 93.33  0.00
```

Thus, we see that locality 21 has precedence with respect to accessibility over 93.33% (42) of the other 45 localities and does not have definite deficiency with regard to any of the other localities. Therefore, the two indicators give mixed messages for the other 6.67% (3) of the localities. Likewise, locality 15 has precedence over 40 of the other 45 localities and has definite deficiency with respect to one of the other localities (locality 21).

3.4 Precedence Plot

To obtain a *precedence plot*, we put the precedence percent on the vertical axis and the definite deficiency percent on the horizontal axis. Function 3.2 is a facility for producing such a plot with some special annotation.

Function 3.2 PrecPlot function for producing a precedence plot in R.

```
PrecPlot <- function(PpDd)
# Input is data frame of rating relations.
# Idz is column number of CaseIDs.
# Pp is column number of PP.
# Dd is column number of DD.
{Cases <- length(PpDd[,1])
 Idz <- 1
 Pp <- 2
 Dd <- 3
 Ymax <- max(PpDd[,Pp])
 Ymin <- min(PpDd[,Pp])
 Xmax <- max(PpDd[,Dd])
 Xmin <- min(PpDd[,Dd])
 Xright <- Ymax
 if(Xmax>Ymax) Xright <- Xmax
 plot(PpDd[,Dd],PpDd[,Pp],ylab="Precedence as %",
```

```
    xlab="Definite deficiency as %",xlim=c(0,Xright))
    YY <- c(Ymax,Ymin)
    XX <- c(100-Ymax,100-Ymin)
    lines(XX,YY,lty=1)
    XX <- c(Xmin,Xmin)
    YY <- c(Ymin,Ymax)
    lines(XX,YY,lty=2)
    XX <- c(Xmin,Ymax)
    if(Xmax>Ymax) XX <- c(Xmin,Xmax)
    YY <- c(Ymin,Ymin)
    lines(XX,YY,lty=2)
    XX <- c(Xmin,100-Ymax)
    YY <- c(Ymax,Ymax)
    lines(XX,YY,lty=2)
    }
```

Function 3.2 is applied to precedence with respect to accessibility as obtained from the product–order protocol to produce Figure 3.4 as follows.

```
> PrecPlot(RdsPrecdn)
```

And some interesting items are then identified.

```
> identify(RdsPrecdn[,3],RdsPrecdn[,2])
```

The points plot within a trapezoid due to the fact that PP, DD, and II total to 100%. Items have greater precedence as height increases on the vertical axis. As points move farther to the right at a given vertical level of precedence, the deficiencies become more definite—thus giving them a more subordinate status. A point can only plot as far right as the downward diagonal

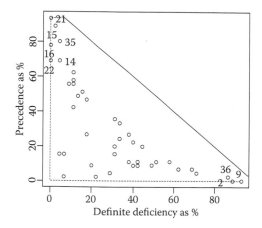

FIGURE 3.4
Precedence plot for access indicators as regional road density and local road density in Lackawanna County with selected points identified.

line, at which position all of the deficiency in precedence is definite with no mixed messages. Again, localities 21 and 15 exhibit the strongest positions of precedence. They are then followed by localities 35, 16, 22, and 14 in that order. Localities 9, 2, and 36 show the least accessibility according to these two indicators.

```
> HighAccess <- c(21,15,35,16,22,14)
> RdsPrecdn[HighAccess,]
      CaseIDs    PP    DD
[1,]       21 93.33  0.00
[2,]       15 88.89  2.22
[3,]       35 80.00  4.44
[4,]       16 77.78  0.00
[5,]       22 68.89  0.00
[6,]       14 68.89  4.44

> LowAccess <- c(9,2,36)
> RdsPrecdn[LowAccess,]
      CaseIDs    PP    DD
[1,]        9  0.00 93.33
[2,]        2  0.00 88.89
[3,]       36  2.22 86.67
```

3.5 Propensities as Progression of Precedence

Precedence plots provide an intuitively interpretable trapezoidal tracking of precedence that is particularly good for finding the best of the best and the worst of the worst. They do, however, have the potential pitfall of masking when it comes to identifying in **R**. If several localities are superimposed, **R** will identify only one of them. More generally, it would also be desirable to have propensities among the instances (localities) revealed in a stated progression of precedence. This can be accomplished through a strategy of ORDering In Tandem (ORDIT).

As PP declines, DD + II increases, and precedence drops. Within deficiency domains of the same size, there is a further sense of handicap as the deficiencies become increasingly definite (more DD relative to II). This suggests ordering as a two-part strategy, with the entire deficiency domain as the first part and the definite deficiencies as the second part. As either of these two parts increases, the instance carries a heavier handicap with regard to precedence. The two parts can be coupled as a decimal value with the entire deficiency domain (call it D) preceding the decimal and the definite deficiency (call it d) following the decimal as a $D.d$ formulation.

Let D be computed by rounding percentage of pairs to two places after the decimal and then multiplying it by 100. Let d be expressed as a decimal fraction of the entire deficiency domain but capped at 0.999 as an upper limit. Expressed in this manner, the two components are simply added together as the ORDIT indicator of progressive precedence, whereby *precedence declines as the indicator increases*. Serial steps of progressively poorer precedence (call them *salient steps*) can be obtained by a regular ranking of the ORDIT indicator, with ties being given the average of the tied ranks. This approach is implemented in ORDITing Function 3.3 for **R**, with the output of the POprecdn function as input. The result is a data frame with Case ID, ORDIT indicator, and salient steps. The salient values are *place ranks* for precedence, whereby rank 1 is the best, and rank 2 is next best. To obtain a progression of precedence reflecting propensity of instances (localities), just reorder the data frame by salient values.

Function 3.3 ORDITing function to compute ORDITs and salient steps.

```
ORDITing <- function(PpDd)
{Ncase <- length(PpDd[,1])
 CaseID <- PpDd[,1]
 ccc <- rep(0,Ncase)
 bbb <- ccc
 ORDIT <- ccc
 for(I in 1:Ncase)
  {ccc[I] <- (100.0 - PpDd[I,2]) * 100.0
   ccc[I] <- round(ccc[I],digits=0)
   if(PpDd[I,2] < 100.0) bbb[I] <- PpDd[I,3]/(100.0 - PpDd[I,2])
   if(bbb[I] > 0.999) bbb[I] <- 0.999
   bbb[I] <- round(bbb[I],digits=3)
   ORDIT[I] <- ccc[I] + bbb[I]
  }
 Salnt <- rank(ORDIT)
 Salnts <- cbind(CaseID,ORDIT,Salnt)
 return(Salnts)
}
```

The ORDITing function is applied as follows.

```
> RdsORDITs <- ORDITing(RdsPrecdn)
> Reordr <- order(RdsORDITs[,3])
> head(RdsORDITs[Reordr,])
       CaseID    ORDIT Salnt
[1,]       21  667.000   1.0
[2,]       15 1111.200   2.0
[3,]       35 2000.222   3.0
[4,]       16 2222.000   4.0
[5,]       22 3111.000   5.0
[6,]       14 3111.143   6.5
> tail(RdsORDITs[Reordr,])
       CaseID      ORDIT Salnt
```

```
[41,]    46  9556.744    41
[42,]     4  9778.068    42
[43,]     1  9778.227    43
[44,]    36  9778.886    44
[45,]     2 10000.889    45
[46,]     9 10000.933    46
```

Case 21 is again seen as having greatest precedence, and case 15 as having next.

3.6 Progression Plot

A *progression plot* can be used to detect ties that cause masking in a precedence plot, and also to identify selected settings in the precedence progression of propensity. Only four **R** commands are needed to generate a progression plot (Figure 3.5) directly as follows.

```
> ORDITrank <- rank(RdsORDITs[,2])
> ORDITstep <- rank(RdsORDITs[,2],ties.method="first")
> plot(ORDITstep,ORDITrank,xlim=c(0,50),ylim=c(0,50))
> identify(ORDITstep,ORDITrank)
```

Tied items will appear horizontally adjacent in a progression plot, whereas there would be masking in a precedence plot. The xlim and ylim specifications can be altered to control what parts of the progression are shown in the plot. Because smaller ORDIT values have more precedence, the precedence is

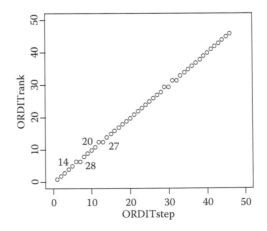

FIGURE 3.5
Progression plot for RdsORDITs with two pairs of ties identified.

greatest at the lower left and least at the upper right. The plot indicates that settings 14 and 28 are tied for the 6th and 7th places, which can be verified by selective retrieval.

```
> RdsORDITs[c(14,28),]
     CaseID    ORDIT Salnt
[1,]     14 3111.143   6.5
[2,]     28 3111.143   6.5
```

3.7 Reversing Ranks

The built-in ranking facility in **R** uses rising ranks, whereby lower rank numbers are less so and higher rank numbers are more so. Place ranks provide a reverse view of ranks as in race ranking, whereby rank number 1 has first place, rank number 2 has second place, and so on. Salient steps have the sense of race ranks (place ranks). It may be appropriate to reverse rankings for the purpose of mapping propensities. This is easily accomplished in **R** simply by adding one to the number of cases and then subtracting the rank vector, which causes **R** to perform the subtraction for each of the rank values. This is demonstrated as follows.

```
> Cases <- length(RdsORDITs[,1])
> FlipSalnt <- (Cases+1) - RdsORDITs[,3]
> RdsORDITS <- cbind(RdsORDITs,FlipSalnt)
> head(RdsORDITS)
     CaseID     ORDIT Salnt FlipSalnt
[1,]      1  9778.227    43         4
[2,]      2 10000.889    45         2
[3,]      3  9111.463    36        11
[4,]      4  9778.068    42         5
[5,]      5  9556.302    40         7
[6,]      6  8000.556    24        23
```

Let us also write this to a text file for bringing back into a geographic information system (GIS) as needed.

```
> write.table(RdsORDITS, "RdsORDITS.txt",row.names=F)
```

Even bringing this table into a GIS can carry seemingly mysterious difficulties. In this situation, the CaseID is the key column that we would expect to use in coupling (joining) this information with that for the locality points as GIS features. The trick is that the ID column for the points has the type of a character string rather than numbers in ArcGIS. We can use Excel© to import the table from **R** into a spreadsheet and specify the string type for

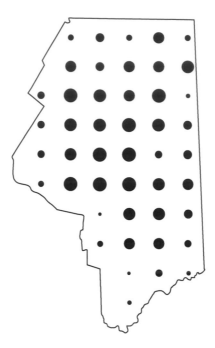

FIGURE 3.6
Map of reversed salient steps as rendered using proportional circular symbols for Lackawanna County so that larger symbols show greater precedence according to road densities.

the CaseID column. The join for the spreadsheet data then goes as expected. Figure 3.6 shows the flipped ORDITs rendered graphically with proportional circular symbols, thus reflecting propensities of localities toward accessibility as seen jointly through regional and local road densities.

3.8 Inconsistency Indicator

The percentage of indefinite instances (II) is useful by itself as an indicator of variability among localities in consistency of the IVIs as comparatives with other localities. Accordingly, it is appropriate to add these to data exported from the statistical software for use in mapping with the GIS.

```
> RdsII <- 100.0 - RdsPrecdn[,2] - RdsPrecdn[,3]
> RdsORDITSII <- cbind(RdsORDITS,RdsII)
> head(RdsORDITSII)
     CaseID     ORDIT Salnt FlipSalnt RdsII
[1,]      1  9778.227    43         4 75.56
[2,]      2 10000.889    45         2 11.11
```

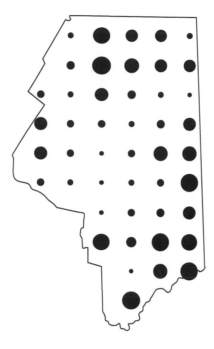

FIGURE 3.7
Differences in inconsistency of road density indicators for Lackawanna County, as reflected in deciles of indefinite instances (II).

```
[3,]      3   9111.463      36        11 48.89
[4,]      4   9778.068      42         5 91.11
[5,]      5   9556.302      40         7 66.67
[6,]      6   8000.556      24        23 35.56
> write.table(RdsORDITSII,"RdsORDSII.txt",row.names=F)
```

The II indefinite instances are then mapped in deciles to reflect differences in inconsistency of indicators for localities as shown in Figure 3.7.

It can be seen that there are large differences in consistency of road indicators, with localities having high road density showing greater consistency (less inconsistency). However, it can be observed by comparison to previous figures that similar situations for roads have similar levels of discrepancy, as would be expected.

3.9 Backdrop: Statistical Software

A major advantage of our localization approach lies in exchange of information between GIS software and statistical software because it provides a row–column data table which is the natural modality of statistical software systems. The exchangeability holds for any statistical software package that accepts and exports delimited American Standard Code for Information Interchange (ASCII) textual files or spreadsheets of data. Spreadsheet systems such as Microsoft Excel will often complement ASCII text files by providing capability for customizing particular aspects of the more generic text file formats.

Statistical software systems mostly fall into one or the other of two general modes of utilization. One mode is via point-and-click dialog boxes that offer choices among several specific options. The other is a command mode that may or may not encompass a programming language of some sort. While the modality of dialog boxes tends to be preferred by those who are less experienced with statistical analysis, the command mode is at least potentially more powerful and adaptable. Because the localization paradigm is inherently flexible and amenable to adaptation, a command-based statistical software system having programmability is advantageous.

The **R** (R Development Core Team 2008) statistical software system has multiple advantages for spatial localization because it is a sophisticated programming language, is available without cost, has an extensive repertoire of packaged protocols, is very extendable by the user community, and has inspired a variety of books dealing with its usage in different contexts. A few such books include Dalgaard (2002), Venables and Smith (2004), Crawley (2005), Everitt (2005), Verzani (2005), Everitt and Hothorn (2006), Murrell (2006), Crawley (2007), Maindonald and Braun (2007), Marques de Sa (2007), Rizzo (2008), Wright and London (2009), Hogan (2010), Keen (2010), Oja (2010), Horton and Kleinman (2011), and Allerhand (2011).

See the Website www.r-project.org as a starting point for obtaining the **R** software by download. The **R** base system is updated quite frequently. There are a great many supplementary packages contributed by users that are available via the Comprehensive R Archive Network (CRAN) with multiple mirror sites around the world. The contributed packages are downloaded individually and must be loaded as libraries to make them active in a session.

R has different types of objects as data models, with the *data frame* being the basic tabular container for cases-by-variables statistical data, wherein cases are rows and variables are columns. The columns of a data frame correspond to the columnar fields of a spreadsheet. A data frame has [row, column] indexing for selective retrieval. A *matrix* object differs in treating all entries as being of the same kind, rather than each column being of a different kind. Each type of object has particular components, with methods appropriate to the context such as for printing.

In its most simplistic personality, **R** can function as a calculator, but more substantial work is done through expressions having strict syntax, with this strictness being one of the more frequent tripwires for casual users. A peculiarity lies in particularizing what would usually be thought of as context for an equal sign (=). In programming languages, the equal sign often serves as an assignment operator by signifying that something on the right side is to be "put into" something on the left side. In **R**, however, this assignment role is more clearly served by a two-character left-pointing arrow (<–) while the "=" sign has other roles such as specifying values for inputs to operational functions. Functional commands have single-word names and are followed by specifications of inputs in parentheses. The command name is actually a name for a *function* facility, with the user having full capability to supply new operations as functions that act as imbedded programs when invoked. A user-created function is loaded from an ASCII text file by the "source("filename")" command.

There is convenient capability via the "File" menu at the top of the desktop to change working directories (folders) and reload system status that was saved at the end of a previous session. If some of a session is to be shown in a document as we do in this chapter, it is most expedient to copy and paste from the **R** desktop into the document.

There is a well-structured but somewhat terse help facility with the help command being "help" and the targeted command of the help being given in parentheses as, for example, "help(help)." The hash sign (#) signifies annotation following on the remainder of the line.

While **R** has quite liberal naming conventions, it is important to avoid what amounts to usurping the name of a regular function by using it for other purposes. For example, using T or F as an object name would interfere with their roles as logical values for TRUE and FALSE. Likewise, a lower-case "t" is the name for the built-in matrix transpose function. Single-character names by naïve users tend to invite difficulty.

References

Allerhand, M. *A Tiny Handbook of R*. New York: Springer, 2011.

Bruggemann, R., and Patil, G. P. Multicriteria prioritization and partial order in environmental sciences. *Environ. Ecol. Stat.* **17**: 383–410, 2010.

Crawley, M. *Statistics: An Introduction Using R*. Chichester, UK: Wiley, 2005.

Crawley, M. *The R Book*. Chichester, UK: Wiley, 2007.

Dalgaard, P. *Introductory Statistics with R*. New York: Springer, 2002.

Everitt, B. *An R and S-PLUS Companion to Multivariate Analysis*. London: Springer-Verlag, 2005.

Everitt, B., and Hothorn, T. *A Handbook of Statistical Analysis Using R*. Chichester, UK: Wiley, 2006.

Hogan, T. *Bare-Bones R: A Brief Introductory Guide*. Los Angeles: Sage, 2010.

Horton, N., and Kleinman, K. *Using R for Data Management, Statistical Analysis, and Graphics*. Boca Raton, FL: Taylor & Francis/CRC, 2011.

Keen, K. *Graphics for Statistics and Data Analysis with R*. Boca Raton, FL: Chapman & Hall/CRC, 2010.

Maindonald, J., and Braun, J. *Data Analysis and Graphics Using R: An Example-Based Approach, 2nd ed.* Cambridge, UK: Cambridge Univ. Press, 2007.

Marques de Sa, J. *Applied Statistics: Using SPSS, STATISTICA, MATLAB and R.* New York: Springer, 2007.

Murrell, P. *R Graphics.* Boca Raton, FL: Chapman & Hall/CRC, 2006.

Myers, W., and Patil, G. P. Preliminary Prioritization Based on Partial Order Theory and R Software for Compositional Complexes in Landscape Ecology, with Applications to Restoration, Remediation, and Enhancement. *Environ. Ecol. Stat.* **17**: 411–436, 2010.

Myers, W. and Patil, G. P. *Multivariate Methods of Representing Relations in R for Prioritization Purposes: Selective Scaling, Comparative Clustering, Collective Criteria and Sequenced Sets,* New York: Springer, 2012.

Oja, H. *Multivariate Nonparametric Methods with R: An Approach Based on Spatial Signs and Ranks.* New York: Springer-Verlag, 2010.

R Development Core Team. *R: A Language and Environment for Statistical Computing.* Vienna, Austria: R Foundation for Statistical Computing, ISBN 3-900051-07-0, URL http://www.R-project.org/2008.

Rizzo, M. *Statistical Computing with R.* Boca Raton, FL: Chapman & Hall/CRC, 2008.

Venables, W., and Smith, D. *An Introduction to R, Revised and Updated.* Bristol, UK: Network Theory Limited, 2004.

Verzani, J. *Using R for Introductory Statistics.* Boca Raton, FL: Chapman & Hall/CRC, 2005.

Wright, D., and London, K. *Modern Regression Techniques Using R: A Practical Guide.* Los Angeles: Sage, 2009.

4

Raster-Referenced Cellular Codings and Map Modeling

4.1 Introduction

As spatial features become small relative to the bounding box, fixed-form features (called *vector data* in geographic information system [GIS] jargon) become less efficient as a means of representing spatial structure. Area declines as the square of spatial span while perimeter declines linearly, giving rise to larger perimeter-to-area ratios and requiring relatively heavier investment in computer storage capacity for explicit coordinates. Likewise, the number of rows in an attribute table becomes so large that direct user access is discouraged. Therefore, a second general mode of managing spatial data becomes more attractive and tractable. It is also conducive to map modeling for synthesizing sophisticated indicators pertaining to complex concepts. This second mode entails *raster referencing* for spatial specification, with *raster* being a row–column pattern of positions.

4.2 Fixed-Frame Micromapping with Conceptual Cells

In the raster realm, the only spatial specifics pertain to a bounding box within which an attribute is specified in small spatial steps of the same size. Position is thus implicitly inferred according to the number of steps taken in moving away from a corner of the bounding box, typically as so many steps "over" and so many steps "down" from the upper-left corner of the bounding box. In this manner, space is effectively seen as a box of virtual boxes forming a two-dimensional array of cells so that computer capacity concentrates on contents of cells rather than coordinates of corners. The content can be quantitative such as concentrations and intensities, or it can be classification codes for qualitative conditions.

Visualizations and cartography for raster map data are constructed by assigning colors to the cell codes and making a paint-by-number picture by coloring in the virtual cell areas according to the numbered palette. Such portrayals are quite effective for depicting fine detail when done with contrasting colors and a limited number of classes. Such displays are particularly good for presentations on computer screens when there is capability for zooming in and zooming out so that overall patterns can be perceived and then interesting areas can be investigated further at a finer spatial scale. Color is costly for print media, however, so economics often dictate that maps be made simpler in suitable shades of gray.

It should be obvious, but bears emphasis, that perception of pattern in a raster map is highly dependent upon the choice of palette for presentation. While the data may be considered objective, mapping as visualization is very much in the subjective realms of interpretation. Given the technical ability of the computer cartographer to alter the palette with ease for any given data domain, there is an inherent story-telling aspect to mapping. Maps thus become tools for persuasive presentation of perspectives (Lang 2001). If the focus is on forests relative to other kinds of land cover, it is expedient to assign the same color to all kinds of forest. Doing so will effectively contrast forests to other things but will eliminate distinctions that are very important to many aspects of landscape ecology. Conversely, the use of strongly contrasting colors for different kinds of forest will tend to suppress perception of more physically obvious differences among other kinds of land cover.

Making a map in color and then reproducing it without color can lead to quite different impressions than originally intended. Likewise, carefully crafted color patterns will mostly escape those in the audience who are colorblind. There is also a somewhat subtle issue around zooming out to get a broader perspective. Rendering the coarse resolution typically involves a dithering process that helps to conceal unwanted detail. Dithering carries an inherent propensity for fuzziness, with this affecting some of the figures in what follows.

4.3 Cover Classes and Localizing Logic

Land cover classification is a major informational theme for raster mapping and serves as the primary topical context for this chapter—specifically, excerpts from the National Land Cover Database (NLCD) produced by the Multi-Resolution Land Characteristics (MRLC) consortium (Chander et al. 2009; Homer et al. 2004).

Each virtual box or cell in raster map data is represented in the data file by a code number for the class which pertains to that particular position. The class codes for the NLCD pertaining to Pennsylvania are given in Table 4.1.

Typically each row of cells accounts for a chain of class codes. The number of cells in a row is specified as part of the header information for the data file along with the number of rows and the size(s) of the row and column steps. The individual cells in the land cover data span 30 meters on a side.

TABLE 4.1

Land Cover Class Codes for NLCD Data Pertaining to Pennsylvania

```
11 = Water
21 = Low intensity residential
22 = High intensity residential
23 = Commercial/industrial/transportation
31 = Barren rock/sand/clay
32 = Quarries, strip mines, gravel pits
33 = Transitional
41 = Deciduous forest
42 = Evergreen forest
43 = Mixed forest
81 = Hay/pasture
82 = Row crops
85 = Urban/recreational grasses
91 = Woody wetlands
92 = Emergent herbaceous wetlands
```

When there are progressions of influences among the classes, these can be cast as lighter or darker shades of gray. In the current case, there is a progression from "naturalistic" forest to moderate human disturbance for grassland to more human disturbance in cultivated agricultural crops to strongly "humanistic" urban residential and commercial/industrial land uses. Figure 4.1 gives this kind of rendering for land cover, whereby forests are dark (lesser human influence) and urban areas are light/bright (stronger human influence). The forested area of Bell Mountain and Griffen Reservoir are at the upper left, and

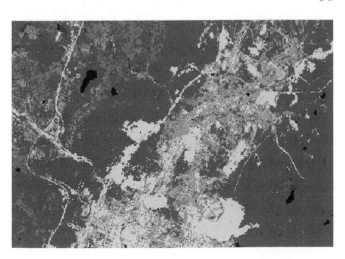

FIGURE 4.1
Gray-shading rendition of land covers with stronger human impact as increasingly lighter tones for the Bell Mountain–Scranton area of Lackawanna County. Water is black, and forest is dark gray, with commercial/industrial and large roads being nearly white.

the heavily urbanized Scranton area is in the lower center. Water is shown as black, forests as dark gray, and commercial/industrial as nearly white.

A more regional view is shown in Figure 4.2 which is intended to show several aspects of advantage in coupling raster and vector modes for different kinds of spatial information. The raster land cover data extend over the whole of Pennsylvania, whereas the focus here is on Lackawanna County. The counties are fairly simple polygons, but it would take a great many more polygons to show all of the very local changes in land cover. A vector county layer has been superimposed on the raster land cover with all of the counties except Lackawanna having opaque shading in the polygons. This creates a window in the area of Lackawanna County, with the shaded counties making a frame through which the land cover is visible. The wider scope incurs a more dithering effect.

Land cover characteristics provide a rich source of indicators regarding the human environment interface. We proceed to consider percentage of developed land (residential, commercial, industrial, transportation) as one example of such an indicator. The relevant NLCD classes are 21, 22, and 23. Accordingly, we need to tabulate the percentage of area in the 2-km proximity perimeters that these three classes collectively comprise. ArcGIS continues to be our "workhorse" GIS in this regard. The Spatial Analyst "extension" includes a facility that provides for tabulating areas by raster class within vector zones, with the output being a special attribute table. After exercising that facility, we augment the table with additional columns and use the "Field Calculator" to obtain the pooled percentage of area among the 20s codes.

As before, we then augment the attribute table for the locality points, join the new table to it, and transfer the values of percent developed land into the newly created column. Figure 4.3 shows the developed land integrated

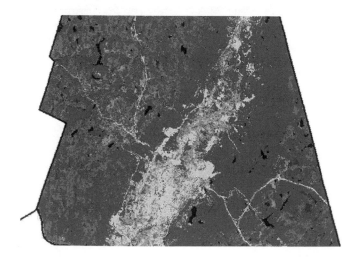

FIGURE 4.2
Extended view of land cover for portion of Lackawanna County showing a window between adjacent counties on left and right sides.

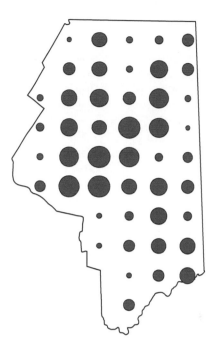

FIGURE 4.3
Decile diagram for percent developed land in Lackawanna County using graduated symbols.

vicinity indicator variate using graduated symbols for deciles. The percentages range from a low of 0.02% to a high of 84.5%, with the first three deciles all being less than 1%. Proportional symbols are not workable in this case because the larger symbols would overlap and cover up their neighbors.

4.4 Raster Regions and Associated Attributes

The simple cell coding of classes outlined above is not the only structure to be found for raster map data. In particular, it may be noted that all attribution resides in the cell codes without provision for indirect attribution as is done in the attribute table for vector data. There is an alternative approach originating with the ArcINFO predecessor to ArcGIS that does incorporate a formal attribute table which is indexed via the codes residing in the cells. This has been given the technical GIS terminology of *grid* structure, which also brings forth strongly the dangers of tangled terminology among various disciplines and subdisciplines in spatial analysis.

One of the crucial concerns in raster representation is the number of different codes that could occur for a cell. If there are 256 or fewer codes (classes), then the information can be contained in one byte of computer data storage per cell in a binary mode. If there are more than 256 possibilities, then two bytes per cell will be required which can accommodate 65,536 different codes. With two bytes per cell, it becomes feasible to consider assigning a different code to each contiguous patch of cells that are of the same kind. Because the code then designates patch rather than class, it becomes necessary to have an attribute table in which to record the class information for the patch. This latter structure has similarities to the vector structure, except that a set of patch codes takes the place of coordinate pairs. Even if patch codes are not used, the indexed attribute table is useful when there are several different but congruent attributes associated with each class as might be the case with the several properties of a soil type. This is especially true when the additional attributes are not specified as simple integer numbers. The raster grid structure mentioned above uses the term *value attribute table* (VAT) for the table of associated attributes.

It is a relatively straightforward matter to generate a raster version of polygonal vector data. One need only step a virtual marker along the center-lines of cell rows while writing the chosen code for the current polygon to an output file. It is a much more difficult proposition, however, to generate a vector version of raster data. This latter basically entails circumscribing connected components among the cells. One approach to accomplishing this is to construct a network of contrasting cell boundaries, blend very small elements into their larger neighbors, and then smooth out the kinky corners of the cells. This tends to be very time consuming and results in very dense vertices along the boundaries, which consumes considerable computer storage capacity and may require "thinning." Given the problematic nature of the raster to vector conversion, it is better avoided insofar as practical.

4.5 Map Modeling

Raster regimes offer a convenient computational context for sophisticated synthesis of innovative indicators. Such scenarios can be considered as *map modeling* (Brimicombe 2010; Liu 2009; Mount et al. 2009). An illustrative scenario should help one to appreciate the possibilities in this regard.

Recall that the individual cells in the land cover data span 30 meters on a side. Suppose that we want to explore influences of development (Shearer et al. 2009) that extend 150 meters from cells that are shown as being developed (i.e., coded as 21, 22, or 23). There is a raster analysis tool in ArcGIS that will create a new raster with selected codes expanded a specified number of cells beyond their current positions. This tool could serve our purpose, but interactions among the three codes would complicate matters. We can avoid these complications by first using a reclass tool to recode the cells so that 21, 22, and 23 are all replaced by 20 as a more generalized code. It will also be convenient to generalize the forest codes so that 41, 42, and 43 all become 40. The land cover data thus generalized will be given the name "NLCDgnrl." Figure 4.4 shows a snippet of the generalized data using gray shades.

Following the recoding, the 20 class of developed areas can be expanded by five surrounding cells. The expanded data corresponding to Figure 4.4 is shown in Figure 4.5. This expansion also has the effect of filling-in the

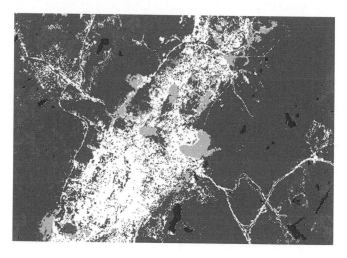

FIGURE 4.4
Snippet of recoded data with all developed classes combined and all forest classes combined. Water is shown as black, forest as dark gray, grass and cropland as light gray, and developed land as white.

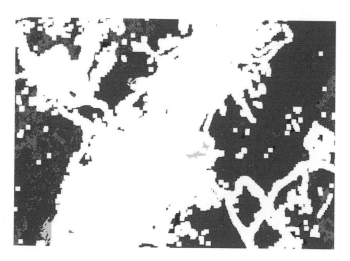

FIGURE 4.5
Snippet of five-cell (150-m) expansion around development. Water is shown as black, forest as dark gray, grass and cropland as light gray, and developed land as white.

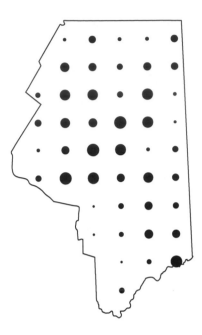

FIGURE 4.6
Decile diagram for modeling of percent developed land with developed land influence in Lackawanna County using graduated symbols.

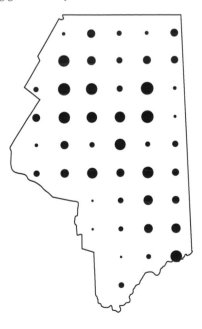

FIGURE 4.7
Decile diagram for modeling of percent influence of developed land in Lackawanna County using graduated symbols.

spaces between spotty developments. The percentage composition of development influence can then be compiled for the vicinities of the localities in the same manner as done prior to recoding and expansion. This is shown in Figure 4.6 using graduated symbols for deciles as was done in Figure 4.3. Due to the nature of the model, the percentages are obviously greater than for developed land itself, ranging from a minimum of 1.9% to 100%. The relative orders in the spatial pattern are similar with some variations.

The influence effect can be obtained by subtraction. This is shown in Figure 4.7, again using graduated symbols for deciles. The minimum of the influence factor is 1.2% and the maximum is 88.2%. The maximum occurs for locality number 4 down in the lower-right corner of the county on the border. This represents a very pronounced fill-in effect of dispersed development. Note that snippets represent only a portion of the county and do not include the lower-right corner.

4.6 Layer Logic

Raster regimes also allow a variety of logical operations to be conducted with raster layers being the operands. Operations of this nature can be used here to support the supposition that the disproportionately large development influence in the vicinity of linkage locale number 4 is due to a fragmental pattern of development.

The chosen approach is to place a condition on cells that show development after the expansion has taken place. Such cells were either in the expansion area or originally coded as development cover. For cells meeting the condition of development after expansion, the code from the generalized land cover layer will be used. For all other cells, the code will be set to zero in the output raster. This effectively masks all cells that were not within the scope of expansion. The output cells having zero as code are then cast in black, while shades of gray are used for other codes with original development cells being white. The result is shown for the lower part of the county in Figure 4.8. It can be seen that the vicinity of locality number 4 does indeed have a fragmental pattern of development that is subject to extensive filling-in.

There is usually more than one operational strategy for accomplishing such an analytical goal in GIS. An alternative in this case would be to make a simple raster mask having the code number 1 for cells showing development after expansion and the code number zero otherwise. In a second operation, "map algebra" would be used to multiply the generalized land covers by the mask on a cell-by-cell basis. The resulting output would again have a zero mask over areas that were not within the scope of expansion.

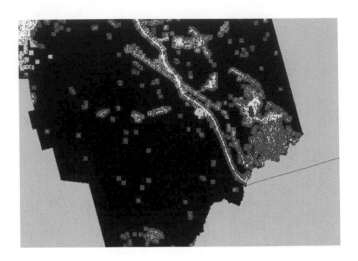

FIGURE 4.8
Masking of cells not within scope of expansion for modeling influence of development. Masked area is black. Dark grays are forest and water. Development is white. Other types are medium to light gray.

Work in this chapter has provided two additional indicators with which to pursue multivariate methods. Localization of the land cover data has given the percentage of developed land, and map modeling has given the percentage of development influence. The compilation of composition also makes the percentage of other land cover classes immediately available.

References

Brimicombe, A. GIS, *Environmental Modeling and Engineering*, 2nd ed. Boca Raton, FL: CRC Press, 2010.

Chander, G., Huang, C., Yang, L., Homer, C., and Larson, C. Developing consistent Landsat data sets for large area applications—the MRLC Protocol. *IEEE Geoscience and Remote Sensing Letters* 6(4): 777–781, 2009.

Homer, C., Huang, C., Yang, L., Wylie, B., and Coan, M. Development of a 2001 National Landcover Database for the United States. *Photogrammetric Engineering and Remote Sensing* 70(7): 829–840, 2004.

Lang, L. *Managing Natural Resources with GIS*. Redlands, CA: ESRI Press, 2001.

Liu, Y. *Modelling Urban Development with Geographical Information Systems and Cellular Automata*. Boca Raton, FL: CRC Press, 2009.

Mount, N., Harvey, G., Aplin, P., and Priestnall, G., Eds. *Representing, Modeling, and Visualizing the Natural Environment*. Boca Raton, FL: CRC Press, 2009.

Shearer, A., Mouat, D., Bassett, S., Binford, M., Johnson, C., Saarinen, J., Gertler, A., and Kahyaoglu-Koracin, J. *Land Use Scenarios: Environmental Consequences of Development*. Boca Raton, FL: CRC Press, 2009.

5

Similar Settings as Clustered Components

5.1 Introduction

This chapter brings together the work of previous chapters in a clustering context to extend explorations of spatial structure in a multivariate mode. We combine percent development (DvlpIVI) and modeled percent of extended development influence (DvlmIVI) from the previous chapter with the two road densities from Chapters 2 and 3 to provide a more substantial suite of indicators, all of which are taken as speaking to some aspect of human imprint on the landscape. We use these indicators for compound clustering to elucidate similar settings. The first stage of our compound clustering serves to extract collectives as "clustered localities agglomerated nonspatially" (CLANs) based entirely on the indicator characteristics without regard to spatial situation. The second stage further segregates the CLANs into "clustered localities using map positions" (CLUMPs) of nearby localities belonging to the same CLAN cluster. We then apply patterns of precedence/propensity as in Chapter 3 and introduce representative ranks of cluster collectives for studying similarity and comparative contrast of spatial settings.

Conventional correlations can provide preliminary perspectives for the clustering context. The Pearson correlation matrix of the four integrative vicinity indicators (IVIs) as obtained from **R** is given in Table 5.1.

Correlations (Pearson) among the development indicators and local roads are fairly strong but not to the degree that would constitute redundancy. As before, there is evident but not particularly strong correlation for regional roads. The corresponding Spearman (rank) correlation matrix is given in Table 5.2.

Whereas some of the Spearman correlations are less than the Pearson correlations, the Spearman correlation of development and modeled development influence is notably greater than the Pearson correlation, with these two being somewhat rank redundant. This is a situation where working entirely with ranks would not tell the full story of relationships in the data. A lattice plot by pairs of the four indicators is shown in Figure 5.1, and a listing of the leading lines for the data frame appears in Table 5.3.

TABLE 5.1

Pearson Correlation Coefficients for Four Indicators

(Pearson)	RrdIVI	LrdIVI	DvlpIVI	DvlmIVI
RrdIVI	1.0000000	0.3739562	0.5054833	0.5983437
LrdIVI	0.3739562	1.0000000	0.8488584	0.7921607
DvlpIVI	0.5054833	0.8488584	1.0000000	0.7841056
DvlmIVI	0.5983437	0.7921607	0.7841056	1.0000000

TABLE 5.2

Spearman Rank Correlation Coefficients for Four Indicators

(Spearman)	RrdIVI	LrdIVI	DvlpIVI	DvlmIVI
RrdIVI	1.0000000	0.3137315	0.6630485	0.6608899
LrdIVI	0.3137315	1.0000000	0.7124884	0.6844897
DvlpIVI	0.6630485	0.7124884	1.0000000	0.9552266
DvlmIVI	0.6608899	0.6844897	0.9552266	1.0000000

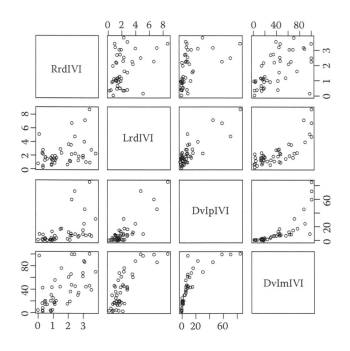

FIGURE 5.1
Lattice of paired plots for regional roads, local roads, developed land cover, and model of developed land influence.

TABLE 5.3

Leading Lines of Data Frame Containing Four Indicators

```
> head(IVI4)
  ID  POINT_X POINT_Y    RrdIVI    LrdIVI   DvlpIVI    DvlmIVI
1  1 450265.1 4561637  0.298855  2.366882  0.995060  14.618082
2  2 450265.1 4566637  0.000000  0.747736  0.028636   2.498568
3  3 455265.1 4566637  1.956432  0.688363  1.911101  14.086321
4  4 460265.1 4566637  0.104275  5.107497  8.684480  96.947584
5  5 445265.1 4571637  0.319571  2.208935  0.035783   1.939454
6  6 450265.1 4571637  1.110692  1.291283  1.001932  13.368639
```

5.2 CLAN Clusters

The first stage of compound clustering is hierarchical clustering (Abonyi and Balaz 2007; Everitt et al. 2001; Kaufman and Rousseeuw 1990; Mirkin 2005; Xu and Wunsch 2009) of localities on indicators using the hclust() facility of **R**. The preliminary to that is calculation of a distance matrix reflecting disparities among the localities.

```
> IVI4dist <- dist(IVI4[,4:7],method="euclidean")
```

A method of linkage among groups to form larger clusters must be chosen for conducting the hierarchical clustering. The "ward" linkage is chosen here due to its tendency to form moderately compact clusters as opposed to "stringy" clusters that wander across the multivariate space of indicators (Podani 2000).

```
> IVI4Hcls <- hclust(IVI4dist,method="ward")
```

The dendrogram of clustering is obtained as follows and shown in Figure 5.2.

```
> plot(IVI4Hcls,labels=F)
```

The heights of the horizontals show the degree of distinctiveness between the clusters. Because hclust() is agglomerative rather than divisive, the later mergers are at the top of the tree. The heights are recorded as a component of the cluster output.

A plot showing the heights working backwards from the last merger is helpful in conjunction with the dendrogram for deciding heuristically how many clusters to retain. A plot of this nature can be obtained as follows and is shown in Figure 5.3.

```
> Nstem <- length(IVI4Hcls$height)
> ClusIVI4hts <- rep(0,Nstem)
> for(I in 1:Nstem) ClusIVI4hts[I] <- IVI4Hcls$height[Nstem - I +1]
> plot(ClusIVI4hts[1:25])
```

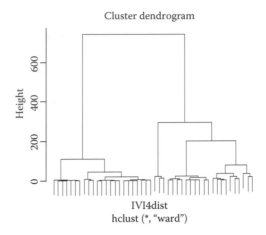

FIGURE 5.2
Dendrogram for hierarchical clustering of localities on four indicators.

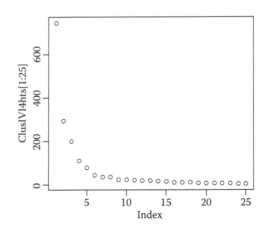

FIGURE 5.3
Plot of heights for horizontals in dendrogram for hierarchical clustering of four indicators with indexing downward in the dendrogram.

The first point in Figure 5.3 accounts for two clusters, and thereafter, each one adds another cluster. When the progression in the figure is examined, it appears that there should be a minimum of three clusters and a maximum of six clusters. Even with only three clusters, one of the clusters will have only three localities. In doing a reconnaissance of the clusters, it will also be helpful to bring forward a copy of Figure 3.3 as Figure 5.4 to show the spatial positions of the members in a cluster.

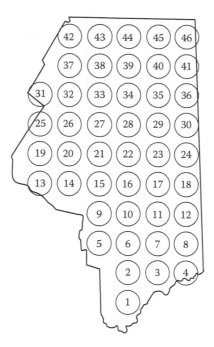

FIGURE 5.4
Map of ID numbers for localities in Lackawanna County.

Cluster membership for a specified number of clusters is obtained with the cutree() command in **R**. We first compare the membership of four clusters and five clusters by cross-tabulation (Myers et al. 2006).

```
> Clus5IVI4 <- cutree(IVI4Hcls,k=5)
> Clus4IVI4 <- cutree(IVI4Hcls,k=4)
> XtabH5xH4 <- table(Clus5IVI4,Clus4IVI4)
> XtabH5xH4
```

```
          Clus4IVI4
Clus5IVI4  1   2   3   4
        1 16   0   0   0
        2  7   0   0   0
        3  0  10   0   0
        4  0   0  10   0
        5  0   0   0   3
```

As can also be seen from the dendrogram, going from four to five clusters entails splitting off 1/3 of cluster 1, which would otherwise contain half the localities. A further split would make another small cluster of three. Therefore, it appears that five clusters are appropriate. The memberships of these five clusters are as follows.

```
> IVI4[Clus5IVI4==1,1]
 [1]  1   3   6 10 12 13 18 23 24 25 31 34 39 44 45 46
> IVI4[Clus5IVI4==2,1]
[1]   2   5   9 19 30 36 42
> IVI4[Clus5IVI4==3,1]
 [1]   4 15 17 22 26 29 32 33 35 37
> IVI4[Clus5IVI4==4,1]
 [1]   7   8 11 16 20 27 38 40 41 43
> IVI4[Clus5IVI4==5,1]
[1] 14 21 28
```

It is interesting to plot each of the four IVIs against cluster number as shown in Figures 5.5 through 5.8. Cluster 1 is low for local roads, development, and

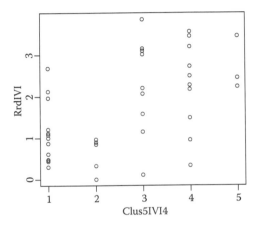

FIGURE 5.5
IVI for regional roads versus CLAN cluster number.

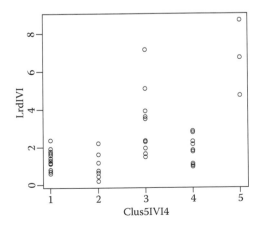

FIGURE 5.6
IVI for local roads versus CLAN cluster number.

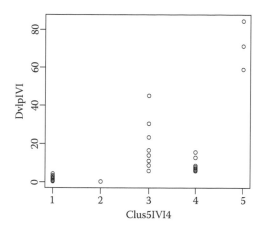

FIGURE 5.7
IVI for developed land versus CLAN cluster number.

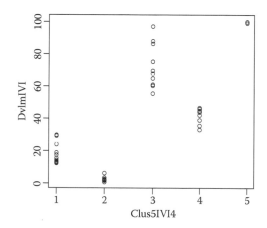

FIGURE 5.8
IVI for modeled development influence versus CLAN cluster number.

development influence but mixed with regard to regional roads. Cluster 2 is low with regard to all four indicators. Cluster 3 is mixed with respect to roads, mostly low with regard to development, but high with respect to development influence. Cluster 4 is mixed with respect to regional roads, low with respect to local roads and development, but moderate with regard to development influence. Cluster 5 is high with respect to all four indicators.

We thus obtain five CLAN clusters based on indicator attributes of localities without explicitly considering spatial position. The CLAN acronym is also suggestive of socialization among smart soft(ware) surveillance sentinels that could be built into advanced agent-based systems.

5.3 CLUMP Clusters

The second stage of compound clustering considers spatial subsets of CLAN clusters that clump together as near neighbors. These could be seen as communities of a CLAN in smart sentinel surveillance scenarios. Thus, we can refer to them as CLUMP community components, or even just as clumps.

Initial inspections followed by clustering coordinates are workable ways of conducting compound clustering. Initial inspection is done by comparing CLAN cluster members as given above with the map positions of localities as shown in Figure 5.4.

CLAN cluster 5 is seen to be small and localized in the middle of the county, with all three of its members being diagonal neighbors and constituting a single CLUMP community. CLAN cluster 1 is geographically distributed, so clustering of coordinates is in order. The "complete" linkage method will be used for this purpose in order to provide very strong resistance to forming stringy clusters.

```
> IVIcls1 <- IVI4[Clus5IVI4==1,]
> IVIcls1xydis <- dist(IVIcls1[,2:3],method="euclidean")
> IVIcls1coms <- hclust(IVIcls1xydis,method="complete")
> plot(IVIcls1coms)
```

The dendrogram for spatial clustering of CLAN cluster 1 is shown in Figure 5.9. It shows four subclusters that are quite well localized. Thus, the CLUMP clusters for CLAN 1 are localities 1, 3, 6, and 10; localities 12, 18, 23, and 24; localities 13, 25, and 31; and localities 34, 39, 44, 45, and 46. Even though the clumps are apparent from the dendrogram, there is still need to extract them formally with a cutree() command.

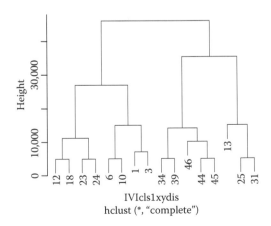

FIGURE 5.9
Dendrogram for CLUMP clustering of CLAN cluster 1.

```
> IVIcls1comn <- cutree(IVIcls1coms,k=4)
> IVIcls1comn
 1   3   6 10 12 13 18 23 24 25 31 34 39 44 45 46
 1   1   1  1  2  3  2  2  2  3  3  4  4  4  4  4
```

Clumping for CLAN 2 is obtained in a similar manner as for CLAN 1, with the dendrogram shown in Figure 5.10.

```
> IVIcls2 <- IVI4[Clus5IVI4==2,]
> IVIcls2xydis <- dist(IVIcls2[,2:3],method="euclidean")
> IVIcls2coms <- hclust(IVIcls2xydis,method="complete")
> plot(IVIcls2coms)
```

Given the isolated nature of localities 19 and 42, there should be four CLUMP community components in CLAN 2, with 19 and 42 being isolates.

```
> IVIcls2comn <- cutree(IVIcls2coms,k=4)
> IVIcls2comn
 2   5   9 19 30 36 42
 1   1   1  2  3  3  4
```

The dendrogram for CLUMP components of CLAN cluster 3 is shown in Figure 5.11 and obtained as follows.

```
> IVIcls3 <- IVI4[Clus5IVI4==3,]
> IVIcls3xydis <- dist(IVIcls3[,2:3],method="euclidean")
> IVIcls3coms <- hclust(IVIcls3xydis,method="complete")
> plot(IVIcls3coms)
```

The dendrogram in Figure 5.11 is indicative of three CLUMP community components with locality 4 being an isolate.

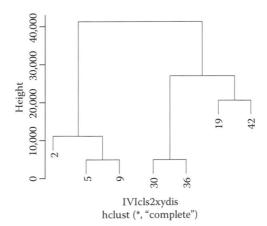

IVIcls2xydis
hclust (*, "complete")

FIGURE 5.10
Dendrogram for CLUMP clustering of CLAN cluster 2.

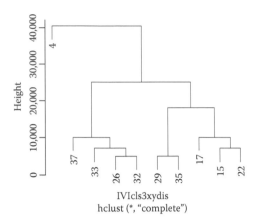

FIGURE 5.11
Dendrogram for CLUMP clustering of CLAN cluster 3.

```
> IVIcls3comn <- cutree(IVIcls3coms,k=3)
> IVIcls3comn
 4 15 17 22 26 29 32 33 35 37
 1  2  2  2  3  2  3  3  2  3
```

The dendrogram for CLUMP community components of CLAN 4 is shown in Figure 5.12 and obtained as follows.

```
> IVIcls4 <- IVI4[Clus5IVI4==4,]
> IVIcls4xydis <- dist(IVIcls4[,2:3],method="euclidean")
> IVIcls4coms <- hclust(IVIcls4xydis,method="complete")
> plot(IVIcls4coms)
```

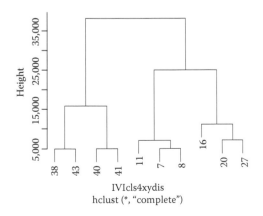

FIGURE 5.12
Dendrogram for CLUMP clustering of CLAN cluster 4.

The dendrogram in Figure 5.12 is not as straightforward as for other CLAN clusters. One CLUMP community should be composed of localities 38 and 43, with another being composed of 40 and 41. However, the branch containing 7, 8, 11, 16, 20, and 27 comprises a somewhat stringy connected component and is appropriately seen as being one community.

Doing a cut with three groups will split the latter set but not the former. Therefore, it is necessary to cut four groups and then editorially recombine two of them.

```
> IVIcls4comn <- cutree(IVIcls4coms,k=4)
> IVIcls4comn
 7  8 11 16 20 27 38 40 41 43
 1  1  1  2  2  2  3  4  4  3
> IVIcls4comn[4:6] <- 1
> IVIcls4comn[8:9] <- 2
> IVIcls4comn
 7  8 11 16 20 27 38 40 41 43
 1  1  1  1  1  1  3  2  2  3
```

The cluster CLANs and CLUMP community memberships can now be added to the locality point file for automated access and mapping. A map showing CLAN numbers for localities appears in Figure 5.13, with

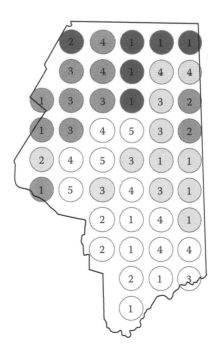

FIGURE 5.13
Map of CLAN cluster numbers for localities with shading for CLUMP communities.

TABLE 5.4

Leading Lines of Data File with CLAN and CLUMP Clusters

ID	POINT_X	POINT_Y	RrdIVI	LrdIVI	DvlpIVI	DvlmIVI	ClanIVI4	Commn
1	450265.1	4561637	0.298855	2.366882	0.995060	14.618082	1	1
2	450265.1	4566637	0.000000	0.747736	0.028636	2.498568	2	1
3	455265.1	4566637	1.956432	0.688363	1.911101	14.086321	1	1
4	460265.1	4566637	0.104275	5.107497	8.684480	96.947584	3	1
5	445265.1	4571637	0.319571	2.208935	0.035783	1.939454	2	1
6	450265.1	4571637	1.110692	1.291283	1.001932	13.368639	1	1

successively darker shading for CLUMP communities within each CLAN. The leading lines of the augmented file are shown in Table 5.4.

5.4 CLAN Cluster Centroids

A conventional way of characterizing groups is to calculate the mean of the group members for each of the variates (indicators), with the resulting vector of values being the centroid of the group. The data frame of centroids for these CLAN clusters is shown in Table 5.5.

We augment the centroids with a column of CLAN cluster numbers, and then make a lattice of paired plots as shown in Figure 5.14. The centroid relations can be compared with the locality relations as shown in Figure 5.1. A precedence analysis like that of Chapter 3 can also be performed on centroids.

TABLE 5.5

Centroids of CLAN Clusters on Four Vicinity Variates

	RrdIVI	LrdIVI	DvlpIVI	DvlmIVI
Clancls1	1.0870601	1.373503	1.393210	18.131609
Clancls2	0.5242067	1.001414	0.055012	3.053203
Clancls3	2.3075421	3.300601	17.163249	72.679581
Clancls4	2.2497409	1.956170	8.410084	41.973361
Clancls5	2.6952030	6.736617	71.502532	99.482232

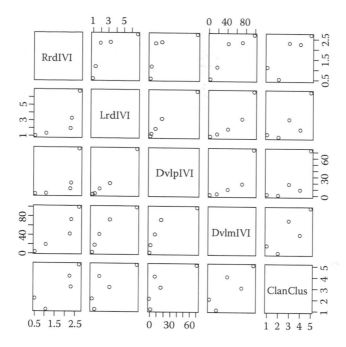

FIGURE 5.14
Lattice plot by pairs for CLAN cluster centroids.

5.5 Salient Centroids

The first step in the precedence analysis is to apply the POprecdn function to the centroids.

```
> BigBetr <- 1
> ClanIDs <- IVI4cntrs[,5]
> CenPrec <- POprecdn(ClanIDs,IVI4cntr,BigBetr)
> CenPrec
      CaseIDs  PP   DD
[1,]        1   25   75
[2,]        2    0  100
[3,]        3   75   25
[4,]        4   50   50
[5,]        5  100    0
```

Next is to make a precedence plot and identify elements as shown in Figure 5.15.

```
> PrecPlot(CenPrec)
> identify(CenPrec[,3],CenPrec[,2])
```

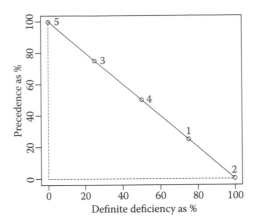

FIGURE 5.15
Precedence plot for centroids of CLAN clusters.

It can be seen from the precedence plot that the CLAN clusters are well ordered with regard to precedence among their centroids. CLAN cluster 5 exhibits the strongest human influence, followed by clusters 3, 4, 1, and then 2 as the least. These relationships can be corroborated by studying Figure 5.14.

We proceed to calculate ORDITs and salient steps and then produce a progression plot as shown in Figure 5.16 (even though masking is absent in this particular situation).

```
> CenORDIT <- ORDITing(CenPrec)
> CenORDIT
      CaseID      ORDIT Salnt
[1,]       1   7500.999      4
```

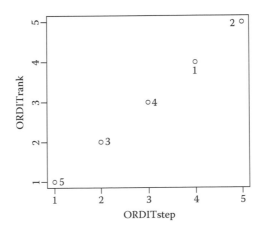

FIGURE 5.16
Progression plot for centroids of CLAN clusters, in which precedence *decreases* from lower left to upper right.

```
[2,]      2 10000.999      5
[3,]      3  2500.999      2
[4,]      4  5000.999      3
[5,]      5     0.000      1

> ORDITrank <- rank(CenORDIT[,2])
> ORDITstep <- rank(CenORDIT[,2],ties.method="first")
> plot(ORDITstep,ORDITrank)
> identify(ORDITstep,ORDITrank)
```

Because the eye is drawn to emphasis rather than diminution in mapping, the salient sequence is more effective when switched in reverse. Henceforth, an "s" will be appended to indicate such switching.

```
> Cases <- length(CenORDIT[,1])
> Salnts <- (Cases+1) - CenORDIT[,3]
> CenORDIT <- cbind(CenORDIT,Salnts)
> CenORDIT
      CaseID      ORDIT Salnt Salnts
[1,]      1  7500.999     4      2
[2,]      2 10000.999     5      1
[3,]      3  2500.999     2      4
[4,]      4  5000.999     3      3
[5,]      5     0.000     1      5
```

5.6 Graded Groups by Representative Ranks

Myers and Patil (2010) present an approach to multivariate comparison of groups that is quite different from centroids. This approach, called *representative ranks*, treats ranks as a kind of common currency among the several indicators.

Representative ranks begin by converting each of the indicators to ranks in the original data matrix from which the groups are drawn. The ranks from all indicators for the members of a group are then collected in a single array, and selected quantiles for the array are determined as the representative ranks. The representative ranks extracted are minimum, first quartile, median, third quartile, and maximum. Function 5.1 is an **R** function for collecting ranks and extracting representatives.

Function 5.1 RepRanks function for extracting representative ranks.

```
RepRanks <- function(Membrshp,CasRnks)
{Groups <- max(Membrshp)
 GroupSiz <- rep(0,Groups)
 Cases <- length(Membrshp)
 Casings <- length(CasRnks)
 for(I in 1:Groups)
```

```
   {J <- Membrshp[I]
    GroupSiz[J] <- GroupSiz[J] + 1
   }
  Min <- rep(0,Groups)
  Q1 <- rep(0,Groups)
  Med <- rep(0,Groups)
  Q3 <- rep(0,Groups)
  Max <- rep(0,Groups)
  RepRnks <- cbind(Min,Q1,Med,Q3,Max)
  for(I in 1:Groups)
   {RnkValus <- GroupSiz[I] * Casings
    RnkList <- rep(0,RnkValus)
    ListIndx <- 0
    for(J in 1:Cases)
     {if(Membrshp[J]==I)
       {for(K in 1:Casings)
         {ListIndx <- ListIndx+1
          RnkList[ListIndx] <- CasRnks[J,K]
         }
       }
     }
    RepRnks[I,1] <- round(min(RnkList),1)
    RepRnks[I,2] <- round(quantile(RnkList,probs=0.25),1)
    RepRnks[I,3] <- round(median(RnkList),1)
    RepRnks[I,4] <- round(quantile(RnkList,probs=0.75),1)
    RepRnks[I,5] <- round(max(RnkList),1)
   }
  RepRnks
}
```

We first proceed to do the rank conversion for the indicator data, and then apply the RepRanks function.

```
> IVI4rnks <- IVI4[,4:7]
> for(I in 1:4)
+ IVI4rnks[,I] <- rank(IVI4rnks[,I],ties.method="average")
> IVI4reprnk <- RepRanks(Clus5IVI4,IVI4rnks)
> IVI4reprnk <- as.data.frame(IVI4reprnk)
> IVI4reprnk
  Min   Q1  Med   Q3 Max
1   3 11.0 16.0 21.0  36
2   1  2.8  4.8  7.2  31
3   2 32.8 38.0 41.0  46
4   6 26.0 30.0 33.5  45
5  32 42.8 44.0 45.2  46
```

5.7 Rank Rods

Rank rods provide convenient visualizations for representative ranks, as shown in Figure 5.17. The total range of ranks is depicted by a vertical line.

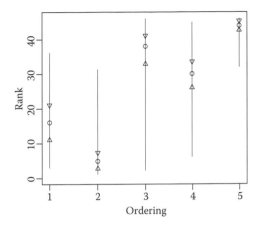

FIGURE 5.17
Rank rods diagram for representative ranks of CLAN clusters. Horizontal ordering is by cluster number. Lower end of each line is minimum rank, and upper end is maximum rank. Upward pointing triangle marks first quartile, and downward pointing triangle marks third quartile. Circle marks median.

The first and third quartiles are shown by triangles appropriately positioned along the line. The median is shown by a circle at the appropriate position on the rod.

Function 5.2 is an **R** facility for plotting rank rods. The first input specifies horizontal ordering of instances (clusters in this case). The second input is the representative ranks, and the third assigns markers to inputs.

Function 5.2 R function for plotting range rods.

```
RankRods <- function(Ordering,RpRnkFram,Ribbing)
{Ribs <- length(Ordering)
 # Ordering is an ordering vector for the ribs.
 # RpRnkFram is a dataframe of representative ranks.
 # Ribbing controls rendering as follows--
 # 0 ignores a column
 # -1 is low end of rib and 1 is high end
 # -2 is upward triangle and 2 is downward triangle
 # 3 is a small circle
 RnkCols <- length(Ribbing)
 Riblo <- 1
 Ribhi <- RnkCols
 LoPnt <- 0
 HiPnt <- 0
 MidPnt <- 0
 for(I in 1:RnkCols)
   {if(Ribbing[I] == -1) Riblo <- I
    if(Ribbing[I] == 1) Ribhi <- I
    if(Ribbing[I] == -2) LoPnt <- I
    if(Ribbing[I] == 2) HiPnt <- I
    if(Ribbing[I] == 3) MidPnt <- I
   }
```

```
MxRnk <- max(RpRnkFram[,Ribhi])
plot(Ordering,RpRnkFram[,Riblo],ylim=c(0,MxRnk),ylab="Rank",pch=" ")
for(I in 1:Ribs)
 {X <- c(Ordering[I],Ordering[I])
  Y <- c(RpRnkFram[I,Riblo],RpRnkFram[I,Ribhi])
  lines(X,Y)
  }
if(HiPnt > 0)
  {for(I in 1:Ribs)  points(Ordering[I],RpRnkFram[I,LoPnt],pch=2)
   for(I in 1:Ribs)  points(Ordering[I],RpRnkFram[I,HiPnt],pch=6)
   }
if(MidPnt > 0)
   for(I in 1:Ribs)  points(Ordering[I],RpRnkFram[I,MidPnt])
 }
```

The RankRods function is applied to representative ranks of CLAN clusters as follows.

```
> ClanID <- c(1,2,3,4,5)
> Rods <- c(-1,-2,3,2,1)
> RankRods(ClanID,IVI4reprnk,Rods)
```

It is apparent from Figure 5.17 that the maxima and minima among groupwise representative ranks are erratic, but the median and quartiles are quite consistent and support the precedence ratings obtained from centroids.

5.8 Salient Sequences by Representative Ranks

The representative ranks can take the role of indicators for plotting precedence and salient sequencing. We pursue the picture presented by rank rods in using the quartiles (including the median as third quartile) as indicators.

In the usual sequencing scenario, we first apply the POprecdn functional facility.

```
> RR3Prec <- POprecdn(ClanIDs,IVI4reprnk[,2:4],BigBetr)
> RR3Prec
     CaseIDs   PP   DD
[1,]       1   25   75
[2,]       2    0  100
[3,]       3   75   25
[4,]       4   50   50
[5,]       5  100    0
```

A precedence plot then follows as shown in Figure 5.18.

```
> PrecPlot(RR3Prec)
> identify(RR3Prec[,3],RR3Prec[,2])
```

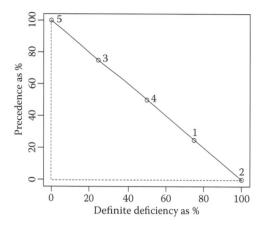

FIGURE 5.18
Precedence plot for Q1, median, Q3 representative ranks of CLAN clusters.

The precedence in Figure 5.18 is consistent with the foregoing, and shows no indefiniteness. The salient sequence is also obtained accordingly.

```
> RR3ORDIT <- ORDITing(RR3Prec)
> RR3ORDIT
       CaseID       ORDIT Salnt
[1,]        1    7500.999     4
[2,]        2   10000.999     5
[3,]        3    2500.999     2
[4,]        4    5000.999     3
[5,]        5       0.000     1
```

The corresponding switched sequence is as follows.

```
> Salnts <- (Cases+1) - RR3ORDIT[,3]
> RR3ORDIT <- cbind(RR3ORDIT,Salnts)
> RR3ORDIT <- as.data.frame(RR3ORDIT)
> RR3ORDIT
  CaseID      ORDIT Salnt Salnts
1      1   7500.999     4      2
2      2  10000.999     5      1
3      3   2500.999     2      4
4      4   5000.999     3      3
5      5      0.000     1      5
```

The preceding analysis can be repeated using all five representative ranks for comparative purposes, giving the precedence plot in Figure 5.19.

```
> RR5Prec <- POprecdn(ClanIDs,IVI4reprnk,BigBetr)
> RR5Prec
```

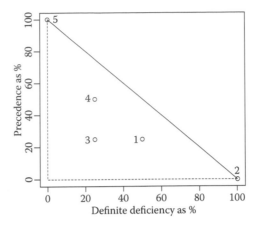

FIGURE 5.19
Precedence plot for minimum, Q1, median, Q3, and maximum representative ranks of CLAN clusters.

```
      CaseIDs  PP   DD
[1,]        1   25   50
[2,]        2    0  100
[3,]        3   25   25
[4,]        4   50   25
[5,]        5  100    0
> PrecPlot(RR5Prec)
> identify(RR5Prec[,3],RR5Prec[,2])
```

The effect of including the more erratic minimum and maximum rank indicators is to give CLAN cluster 4 greater precedence than 3, and to move 1, 4, and 3 to the left of the limiting line showing a greater number of indefinite instances. Number 5 still has definitely greatest precedence, and number 2 still has definitely least precedence. The salient sequencing is as follows.

```
> RR5ORDIT <- ORDITing(RR5Prec)
> RR5ORDIT
      CaseID      ORDIT Salnt
[1,]       1   7500.667     4
[2,]       2  10000.999     5
[3,]       3   7500.333     3
[4,]       4   5000.500     2
[5,]       5      0.000     1
```

The switched salient sequence for five representative ranks is then obtained.

```
> Salnts <- (Cases+1) - RR5ORDIT[,3]
> RR5ORDIT <- cbind(RR5ORDIT,Salnts)
> RR5ORDIT <- as.data.frame(RR5ORDIT)
```

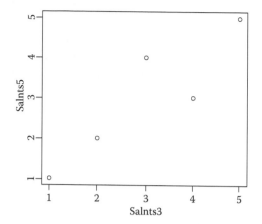

FIGURE 5.20

Plot of switched salient sequence for five representative ranks versus switched salient sequence for three representative ranks.

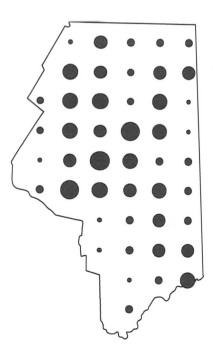

FIGURE 5.21

Map of precedence (switched salient) for CLAN clusters with graduated symbols (largest has greatest precedence).

```
> RR5ORDIT
  CaseID      ORDIT Salnt Salnts
1       1  7500.667     4      2
2       2 10000.999     5      1
3       3  7500.333     3      3
4       4  5000.500     2      4
5       5     0.000     1      5
```

The switched salient sequences for three and five representative ranks are plotted against each other in Figure 5.20. The exchange of roles for CLAN clusters 3 and 4 is evident in Figure 5.20. Because the results for three representative ranks match those for centroids, this salient sequencing will be favored here.

Figure 5.21 is a map of precedence (switched salient) for CLAN clusters.

References

Abonyi, J., and Balaz, F. *Cluster Analysis for Data Mining and System Identification*. Berlin: Birkhauser, 2007.

Everitt, B., Landau, S., and Leese, M. *Cluster Analysis*. London: Arnold, 2001.

Kaufman, L., and Rousseeuw, P. *Finding Groups in Data: An Introduction to Cluster Analysis*. New York: Wiley, 1990.

Mirkin, B. *Clustering for Data Mining: A Data Recovery Approach*. Boca Raton, FL: Chapman & Hall/CRC, 2005.

Myers, W., McKenny-Easterling, M., Hychka, K., Griscom, B., Bishop, J., Bayard, A., Rocco, G., Brooks, R., Constantz, G., Patil, G. P., and Taillie, C. Contextual clustering for configuring collaborative conservation of watersheds in the mid-Atlantic highlands. *Environmental and Ecological Statistics* 13(4): 391–407, 2006.

Myers, W., and Patil, G. Preliminary prioritization based on partial order theory and R software for compositional complexes in landscape ecology, with applications to restoration, remediation, and enhancement. *Environmental and Ecological Statistics* 17: 411–436, 2010.

Podani, J. *Introduction to the Exploration of Multivariate Biological Data*. Leiden, The Netherlands: Backhuys Publishers, 2000.

Xu, R., and Wunsch, D. *Clustering*. New York: Wiley, 2009.

6

Intensity Images and Map Multimodels

6.1 Introduction

The topic of raster-referenced data in Chapter 4 has focused mostly on classified maps such as land cover, with a moderate map modeling being conducted. Other types of data can be represented in rasters, however, especially as intensity images. The general structure of intensity images is like that of cellular class codes, but numbers contained in the virtual cells are quantitatively varying values rather than category codes. The quantities may be concentrations of chemicals, temperatures, rainfall, elevations, slopes of surfaces, distances to particular places like streams or supply sources, or intensity of a signal such as remotely sensed spectral bands. Distances give rise to interposed distance indicators (IDIs). Integrated vicinity indicators (IVIs) may be statistics of strength or vicinity variability. Likewise, modelings with such data often involve the interplay of several models working in complementary combinations (Ford 2009; Parker and Asensio 2008; Skidmore 2002). In this chapter, we explore some of these more general contexts, and we close the chapter with a backdrop on pixelized pictures and remote sensing.

6.2 Intensity as Frequency of Occurrence

A map model was used previously for presentational purposes in Chapter 4. A somewhat more sophisticated scenario will serve here for introducing intensity image information. Making map masks is again a major method for the modeling. This model will seek to show areas of possible interest as exurban residential sites with some wooded ambiance. Not wishing to fragment fully forested areas, however, our first criterion is that there is already at least 20% low-density residential land cover within a 450-m radius of a possible position. The second criterion is that there is at least 30% forest within that same range. Thus, low-density residential and forest will account for more than 50% of the land cover. The third criterion is that not more than 15%

of land cover be commercial/industrial/transportation in that same span of distance.

We begin with the raster map of land cover described earlier as a base layer. We begin to address our first criterion by making a mask (1 = present or 0 = absent) showing the presence of low-density residential land cover (class 21). This mask is made by reclassifying the codes into a new layer so that code 21 is replaced by 1 and all other codes are replaced by 0. We then exercise a geographic information system (GIS) facility that sums the values of cells in a neighborhood. Because each cell spans 30 m, a 15-cell radius is appropriate for this purpose. Inasmuch as the mask contains only values of 0 or 1, summation gives a count of low-density residential cells, thus creating an intensity image for low-density residential in the neighborhood. Also, the 450-m radius encompasses 706 cells. Thus, a sum (count) of 141 cells represents 20% of the composition for the specified neighborhood. A portion of the low-density residential intensity image is shown in Figure 6.1 with darker shades indicating greater intensity.

A conditional operation is next used to create another mask having a value of 1 for intensities greater than 141 or a value of 0 for intensities of 141 or less. This latter mask thus shows where the first criterion is satisfied. We thus go from a mask through an intensity image and then back to a mask. The low-density criterion mask image is shown in Figure 6.2.

Another use for the criterion mask is to impose a threshold of intensity on the intensity image. This can be accomplished with "image algebra," whereby the value of each cell in the intensity image is multiplied by the value of the mask for that same cell position. Thus, all intensity cells having values of 141

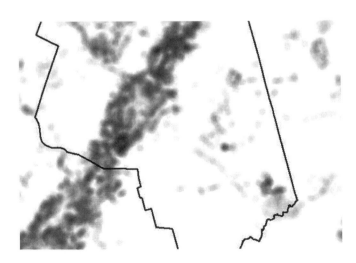

FIGURE 6.1
Intensity image for low-density residential development within a distance of 15 cells from focal cell. Darker shades indicate greater intensity.

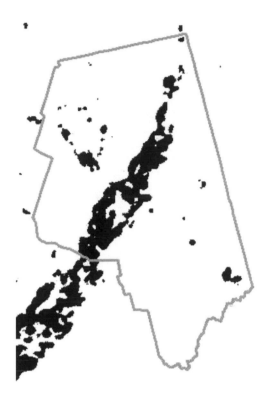

FIGURE 6.2
Low-density (20% or more in a 450-m radius) criterion image mask.

or less will take on the value 0. The threshold could also be imposed as a conditional, or by setting the symbols so that lower values are not evident. These exemplify the frequent availability of several ways for achieving the same effect. A portion of the threshold image is shown in Figure 6.3 as successive steps of darker shading giving the visual impression of contours.

The second criterion of percentage forest cover is addressed in a similar manner but proceeds from the generalized land cover data in which all three types of forest have been reclassified to have the common code of 40. The first step is again to make a mask showing presence of forest as 1 and absence of forest as 0. This is followed by the neighborhood summation operation producing the forest intensity image as partially depicted in Figure 6.4.

The threshold marking 30% forest cover in circular zones is considered here as being 212 or more (greater than 211). The criterion mask for this threshold is shown in Figure 6.5.

It is interesting to see what areas would satisfy the first two criteria before imposing the third. Accordingly, map algebra is performed on the first two criterion masks by adding them together. Thus, a cell value of 2 indicates areas

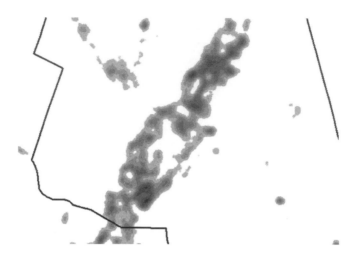

FIGURE 6.3
Portion of intensity image for low-density residential development above threshold of 141 cells within a 15-cell (450-m) radius.

FIGURE 6.4
Intensity image for forest cover within a distance of 15 cells from focal cell. Darker shades indicate greater intensity.

that satisfy the low-density residential and forest cover criteria. Figure 6.6 shows these areas in gray, with county boundary and locality points in black.

The third criterion regarding commercial/industrial/transportation starts with a mask showing land cover class 23 as presence = 1 and absence = 0. The neighborhood summation operation producing the intensity image is partially

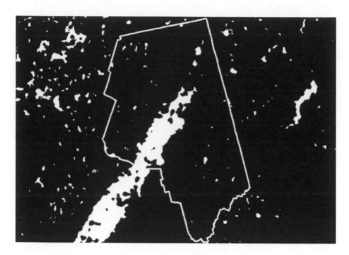

FIGURE 6.5
Forest (30% or more in a 450-m radius) criterion image mask.

FIGURE 6.6
Areas satisfying both the low-density residential and forest cover criteria.

depicted in Figure 6.7 showing the strong influence of roads. The criterion mask for this class comprising 15% or less of land cover is shown is Figure 6.8.

The low commercial/industrial/transportation third criterion is combined with the other two (see Figure 6.6) by multiplying the summation mask of the first two by the mask for the third. This will change to zero anything in the summation mask that does not meet the third criterion. The result for

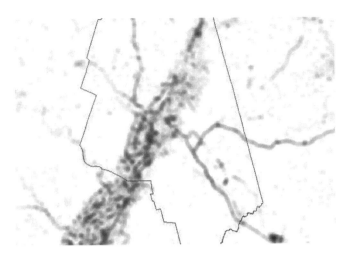

FIGURE 6.7
Portion of intensity image for commercial/industrial/transportation within a distance of 15 cells from focal cell. Darker shades indicate greater intensity.

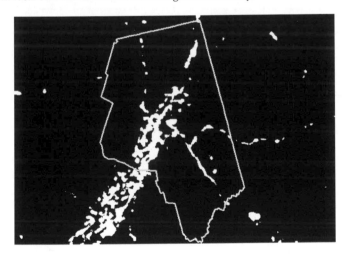

FIGURE 6.8
Criterion image mask for commercial/industrial/transportation comprising 15% or less of land cover (black = true, white = false).

all three criteria is shown in Figure 6.9 with county outline, locality points, and 2-km circles. It is apparent why one would want the full GIS layer for actual logistics of searching for suitable sites. Several of the circular zones are devoid of suitable sites.

Integrated vicinity indicators complement GIS layers. An XurbnIVI is compiled as percent occurrence of site suitability in proximity perimeter. This is shown using proportional symbols in Figure 6.10. The difference in orientation of the county between Figure 6.9 and Figure 6.10 is due to the raster data

FIGURE 6.9
Areas satisfying all three site criteria, showing localities and 2-km (radius) circles.

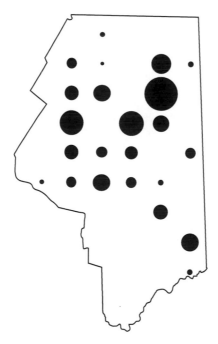

FIGURE 6.10
XurbnIVI for percentage site suitability as shown using proportional symbols.

having a different projection than the vector data. In Figure 6.9, the vector data are dynamically reprojected to match georeferencing of the raster.

6.3 Hillshades and Slopes

Intensity images can be treated in a general sense as topographies. In so doing, there are several GIS facilities for topographies that become useful for visualization and local variation. Virtual hillshading is helpful for visualization. As a case in point, consider the intensity of low-density residential development that provided the first criterion in the previous section. The proximity perimeters for localities are shown on hillshading in Figure 6.11. A vertical exaggeration factor of three has been built into the generation of the hillshading backdrop.

Hillshading is visually effective, but it is influenced by the position of a virtual sun as well as height and steepness of the virtual topography. Steepness of the virtual topography is measured by slope as the change in height when moving from a cell to a neighboring cell. For actual topographies, slope is typically expressed as vertical change as a percent of horizontal change. Topographic facilities for calculating slope are important components of GIS. An (intensity) image of slope for the frequency of low-density residential is shown in Figure 6.12.

Variability of intensity for an image as a topography is important complementary information to the height (intensity) of the topography itself. Cell to cell slope is the most local reflection of such variability available. Figure 6.13

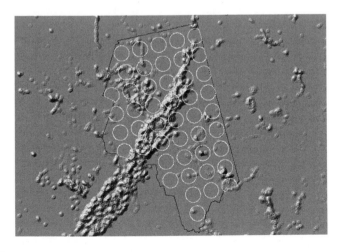

FIGURE 6.11
Proximity perimeters for localities on a hillshade backdrop of intensity for low-density residential development with a threefold vertical exaggeration.

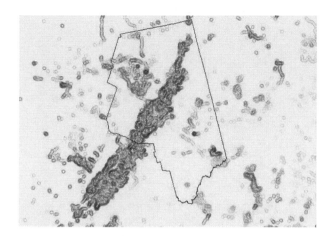

FIGURE 6.12
Intensity image for slope of low-density residential frequency.

is a decile map showing an IVI of mean slope for frequency of low-density residential.

Standard deviation of frequency for low-density residential can serve as a broader-range IVI of variability that does not involve contiguity. Deciles for this IVI are shown in Figure 6.14, with the pattern being very similar to Figure 6.13.

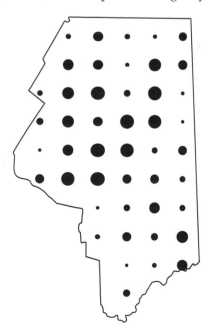

FIGURE 6.13
Deciles of mean slope as IVI for local variability in frequency of low-density residential corresponding to Figure 6.12.

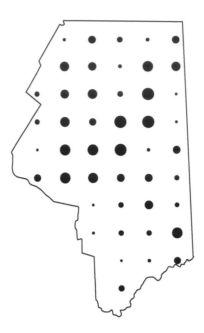

FIGURE 6.14
Deciles of standard deviation for frequency of low-density residential.

Actual topographies are most commonly represented as digital elevation models with each virtual cell containing the elevation of its center point (Maune 2007). For such topographies, *aspect* (as direction of steepest slope) and curvatures are also important.

6.4 Interposed Distance Indicators

Interposed distance indicators have had several mentions but no previous exemplification. We proceed to make such a construct for distance to water in the land cover map. This begins by making a water mask, but this time treating absence of water as "NoData." This mask serves as the input for obtaining distance to water with a GIS analysis facility. Figure 6.15 is a gray-shaded map of the resulting distance to water with darker tones indicating greater distance. In several senses, greater distance to water is associated with lesser ecological sensitivity for this region.

Mean distance for the vicinity of each locality can be used as an IDI for the water context. This is shown as deciles in Figure 6.16.

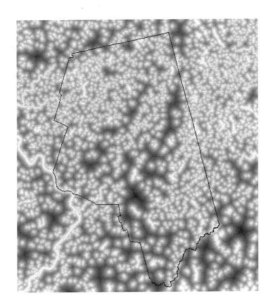

FIGURE 6.15
Intensity image of distance (darker) to water element of land cover.

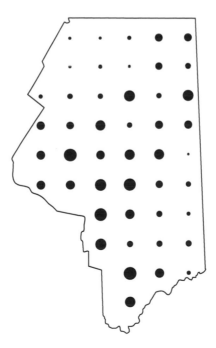

FIGURE 6.16
Deciles of mean distance to water as IDI.

6.5 Backdrop: Pictures as Pixels and Remote Sensing

A special genre of intensity images produces perceptions of pictures. These have become part of everyday experience through familiarity with digital cameras and Google Earth©. Herein the raster frame becomes a "scene" and the raster cell becomes a "pixel," which is short for "picture element." The scene is illuminated by some source of radiant energy such as the sun, and the radiant energy is reflected from the scene to a sensor such as a digital camera. The more energy received in the picture elements, the brighter that portion of the image. Such pictures are thus very fine mosaics of pixels, with a megapixel being one million pixels. When comprising a view of earth from above, scenes can in turn be edge-matched and assembled as blocks in a higher-order mosaic. If the mosaic of blocks comprises too much information to be conveniently handled as a unit in the computer, then blocks of blocks can be treated as tiles of image information in a still higher-order mosaic.

Figure 6.17 shows one scene (or picture) from the PAMAP (PA MAPping) program of digital aerial photography near locality number 34 in Lackawanna County, with the 2-km radius vicinity circles superimposed. It should be apparent that such imagery can be used in a visual interpretation mode to make ocular estimates ("estimetric" indicators) regarding conditions of interest that may not be otherwise available.

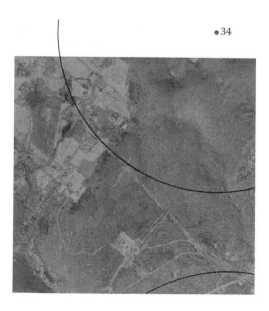

FIGURE 6.17
One scene or picture from PAMAP digital aerial imagery near locality number 34, with vicinity circles having a 2-km radius superimposed.

Things in the earth environment have different reflective responses to qualities of light, which actually arise from the complementary absorptive action. Thus, healthy vegetation selectively absorbs light in broadly blue and broadly red regions of the spectrum of wavelengths. Predominantly green light thus remains to be reflected from leaves, consequently giving a green appearance. Therefore, contrary to popular wisdom, green is the color primarily rejected by plant life. A plot of percentage reflectance versus wavelength from violet to red and beyond constitutes a spectral signature for a substance by which it can be identified, much like a fingerprint.

The human eye has sensory elements that also respond selectively to the broadly blue, broadly green, and broadly red regions of the spectrum (RGB response). Therefore, creating colored images requires making three separate intensity images from the corresponding spectral regions. Rendering tricolor composite intermixed images on the display screen in red, green, and blue tones will produce perception of a color scene. Therefore, presenting a colored image requires three times the computer capacity as for a single gray-tone panchromatic (across the colors) image.

When we analytically address color contrasts, we enter the realms of *remote sensing* science (Campbell and Wynne 2011; Cracknell and Hayes 2007; Jensen 2007; Myers and Patil 2006). However, remote sensing is not restricted to color contrasts.

Special sensors can record regions of the electromagnetic spectrum that are invisible to the human eye. Even extending a little beyond the red into the infrared region can be very revealing. Healthy vegetation is very reflective (bright) in this near (to the visible) infrared region, and water is very absorptive (dark).

Greater spectral selectivity and sensitivity are obtained by using scanning sensors instead of cameras. Cameras are framing sensors that open a shutter and collect data for all pixels essentially simultaneously. A scanner uses moving instantaneous field(s) of view (IFOV), with an IFOV collecting data for one pixel at a time. The pixel-by-pixel information is then assembled into a collective scene, often with many more than three spectral sectors (bands). Scanners are also well suited to operation from satellites (Chuvieco and Huete 2010).

The multiband spectral data are processed with special image analysis software (Tso and Mather 2009). MultiSpec© (engineering.purdue.edu/~biehl/MultiSpec/) is an entry-level version of such software that is publicly available from the LARS laboratory at Purdue University. An excellent tutorial on multiband remote sensing composed by Nicholas Short is available through NASA's Goddard Space Flight Center at the rst.gsfc.nasa.gov Website.

An excerpt from a scene encompassing the Wyoming Valley collected by NASA's Landsat MSS sensor is shown for an infrared band (0.8–1.1 μm) in Figure 6.18 as rendered with MultiSpec. Note the darkness of the water and the urbanized valley contrasting with the brightness of the forested uplands. A MultiSpec rendering of the red band (0.6–0.7 μm) for the same area is

FIGURE 6.18
Near-infrared band (0.8–1.1 μm) for an excerpt from a Landsat MSS scene encompassing the Wyoming Valley as rendered with MultiSpec software.

shown in Figure 6.19 for comparison. Note in the latter case that the forested uplands appear dark rather than bright, whereas the urbanized area and scattered clouds appear bright. There is also a defect in the image data for Figure 6.19 that creates a white line as an artifact.

Distinctive combinations of reflectance in different bands allow automated mapping of land cover components by multivariate statistical discrimination techniques. "Training set" samples of the respective land cover components

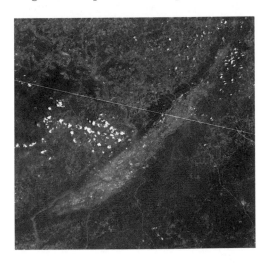

FIGURE 6.19
Red band (0.6–0.7 μm) for an excerpt from a Landsat MSS scene encompassing the Wyoming Valley as rendered with MultiSpec software.

are located in the scene from which the statistics for spectral signatures are extracted. The remainder of the scene is then subjected to statistical classification by pattern recognition protocols (Jones and Vaughn 2010; Lillesand et al. 2008). Thus, the land cover is obtained from spectral/statistical models and is subsequently used for map models.

References

Campbell, J., and Wynne, R. *Introduction to Remote Sensing, 5th ed.* New York: The Guilford Press, 2011.

Chuvieco, E., and Huete, A., Eds. *Fundamentals of Satellite Remote Sensing.* Boca Raton, FL: Taylor & Francis/CRC, 2010.

Cracknell, A., and Hayes, L. *Introduction to Remote Sensing, 2nd ed.* Boca Raton, FL: CRC Press, 2007.

Ford, A. *Modeling the Environment, 2nd ed.* Washington, DC: Island Press, 2009.

Jensen, J. *Remote Sensing of the Environment: An Earth Resource Perspective.* Upper Saddle River, NJ: Pearson Prentice-Hall, 2007.

Jones, H., and Vaughn, R. *Remote Sensing of Vegetation: Principles, Techniques and Applications.* Oxford, NY: Oxford Univ. Press, 2010.

Lillesand, T., Kiefer, R., and Chipman, J. *Remote Sensing and Image Interpretation.* New York: Wiley, 2008.

Maune, D., Ed. *Digital Elevation Model Technologies and Applications: The DEM Users Manual.* Bethesda, MD: American Society for Photogrammetry and Remote Sensing, 2007.

Myers, W., and Patil, G. P. *Pattern-Based Compression of Multi-Band Image Data for Landscape Analysis.* New York: Springer, 2006.

Parker, A., and Asensio, E. *GIS and Spatial Analysis for the Social Sciences: Coding, Mapping, and Modeling.* New York: Routledge, 2008.

Skidmore, A., Ed. *Environmental Modeling with GIS and Remote Sensing.* New York: Routledge, 2002.

Tso, B., and Mather, P. *Classification Methods for Remotely Sensed Data.* Boca Raton, FL: CRC Press, 2009.

7

High Spots, Hot Spots, and Scan Statistics

7.1 Introduction

We are often interested in places where some property is particularly high, which we will call *high spots*. We will use the percentage of area having intensive development of the commercial/industrial/transportation (CIT) type as an exemplar. One way is to proceed in the manner of previous chapters to map this percentage in terms of deciles as shown in Figure 7.1. One is then able to see structure of high spots and act accordingly. This *"see structure =>act accordingly"* paradigm is semistatistical by virtue of having heuristic endpoints regarding the structure that is "seen." Whether something is "high" or not still depends on *declaratively deciding* some threshold. Quantitative characteristics furnish a foundation for declaratively deciding, but the decision has not been based on a statistical scenario of probabilistic protocols for hypothesis testing.

With disease dynamics among major motivations, there has been substantial statistical accomplishment in this arena accompanying exponential evolution in computing capability (Lawson and Denison 2002). Scan statistics have been particularly productive in these regards. Background on the development of scan statistics is provided in the opening chapter of a book dedicated to Joseph Naus as the "father of scan statistics" (Wallenstein 2009).

The general question that is addressed in the current context is whether there are contiguous clusters in space having an elevated risk of being affected as compared to other elements in some spatial extent. Detection is done by scanning a window of some subspace over the entire extent, looking for the most likely subspace(s) and then using Monte Carlo simulation strategies to assess the probability of observing as much or more contiguous concentration under a null hypothesis of a designated distribution. If the simulations show statistical significance, the contiguous cluster(s) are considered to be "hotspots" as opposed to just "high spots." A hotspot is not a matter of a simple threshold but is more of a hill form. A relatively lower hill in a generally low-lying area can constitute a hotspot. Thus, there is effectively a contrasting of a prospective hotspot to its spatial

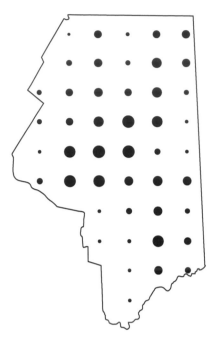

FIGURE 7.1
Percentage of area covered by CIT development mapped in terms of deciles for 2-km radius proximities in Lackawanna County.

surroundings. Time periods can also be added as an additional dimension to spatial extents.

7.2 SaTScan™

SaTScan (http://www.satscan.org/) is a very accessible facility by Martin Kulldorff (1997) that is available along with extensive lists of references and examples without cost for detecting spatial and spatial–temporal clusters under several different distributional assumptions. We apply that facility here to give a flavor of the scenarios with scan statistics. This facility has a documented history of application in several contexts besides mapping of medical maladies (Costa and Kulldorff 2009). Our particular matching of methodologies will make the occurrence of CIT development an analog of a medical malady. SaTScan accepts a set of points as surrogates for spatial sectors, with point proximity as the contiguity condition. This fits well with our scenario of surveillance settings and vicinity variates. SaTScan uses progressively expanding circles or ellipses as the scanning windows that capture the points. The method is motivated by deciding on a distributional model to obtain a null hypothesis. We will use a binomial model and a Poisson model to ask different questions of the data.

7.3 Concentration of CIT Core Development

We first use a binomial model to find hotspots of CIT core development. The percentages of CIT development as mapped in Figure 7.1 are scaled up to 10,000 pixels as integers for specification of cases and controls. The case component is entered in SaTScan's "case" file, and the noncase component is specified in a "controls" file. The point coordinates appear in the geographic file. The ID number of the locality is the first item on each line of all three files, whereby the information is matched across input files. The output of the analysis is as follows.

```
                        SaTScan v9.1.1


Program run on: Sat Aug 06 14:24:56 2011

Purely Spatial analysis
scanning for clusters with high rates
using the Bernoulli model.
_____

SUMMARY OF DATA

Study period..................: 2000/1/1 to 2000/12/31
Number of locations...........: 46
Total population..............: 4600000
Total number of cases.........: 162281
_____

MOST LIKELY CLUSTER

1.Location IDs included.: 21, 20, 15, 27, 22, 14, 26, 16, 28
   Coordinates / radius..: (445265,4.58664e+006) / 7071.07
   Population............: 900000
   Number of cases.......: 100815
   Expected cases........: 31750.63
   Observed / expected...: 3.18
   Relative risk.........: 6.74
   Log likelihood ratio..: 73674.305650
   P-value...............: < 0.000000000000000010
```

```
SECONDARY CLUSTERS


2.Location IDs included: 40
   Coordinates / radius..: (455265,4.60164e+006) / 0
   Population............: 100000
   Number of cases.......: 4932
   Expected cases........: 3527.85
   Observed / expected...: 1.40
   Relative risk.........: 1.41
   Log likelihood ratio..: 265.070769
   P-value...............: < 0.000000000000000010
```

```
PARAMETER SETTINGS

Input
-----
   Case File          : D:\satscans\CITcases.cas
   Control File       : D:\satscans\CITctl.txt
   Coordinates File   : D:\satscans\Settings.geo
   Time Precision     : None
   Start Time         : 2000/1/1
   End Time           : 2000/12/31
   Coordinates        : Cartesian

Analysis
--------
   Type of Analysis   : Purely Spatial
   Probability Model  : Bernoulli
   Scan for Areas with : High Rates

Output
------
   Results File : CIThibin
   Cluster File : D:\Program Files\SaTScan\CIThibin.col.txt

Data Checking
-------------
   Temporal Data Check      : Check to ensure that all cases
and controls are within the specified temporal study period.
   Geographical Data Check  : Check to ensure that all
observations (cases, controls and populations) are within the
specified geographical area.

Spatial Neighbors
-----------------
   Use Non-Euclidian Neighbors file  : No
   Use Meta Locations File           : No
```

```
    Multiple Coordinates Type          : Allow only set of
coordinates per location ID.

Spatial Window
--------------
    Maximum Spatial Cluster Size   : 50 percent of population at
risk
    Window Shape                   : Circular
    Isotonic Scan                  : No

Inference
---------
    P-Value Reporting                  : Default Combination
    Adjusting for More Likely Clusters : No
    Number of Replications             : 999

Clusters Reported
-----------------
    Criteria for Reporting Secondary Clusters : No Geographical
Overlap

Additional Output
-----------------
    Report Critical Values     : No
    Report Monte Carlo Rank    : No
    Print ASCII Column Headers : No

Run Options
-----------
    Processer Usage    : All Available Proccessors
    Logging Analysis   : Yes
    Suppress Warnings  : No
```

```
Program completed    : Sat Aug 06 14:27:25 2011
Total Running Time   : 2 minutes 29 seconds
Processor Usage      : 2 processors
```

It is important to note that the temporal time span specifications are there just as fillers when the "Time precision" is set to "None." SaTScan has found two clusters, with the primary cluster composed of nine localities and a secondary single cluster (locality 40). Both clusters are significant with very low probability. The clusters are shown on "spots" at the settings in Figure 7.2. Comparing these to the decile map in Figure 7.1, one might have anticipated that setting 29 would also be in the cluster. However, it must be remembered that the scanning window is circular. Expanding the window would have also picked up lesser settings along with number 29. Thus, SaTScan does not favor irregularly shaped clusters.

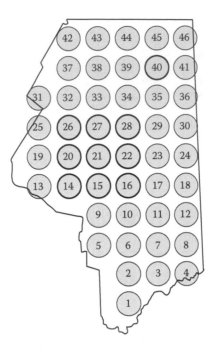

FIGURE 7.2
Significant clusters (hotspots) for concentration of CIT core development by the binomial (Bernoulli) model in SaTScan.

7.4 Complexion of CIT Developments

We change the model to explore a different aspect of places having CIT development. We use our development influence model from Chapter 4 along with the Poisson model of SaTScan to see how CIT development affects its surroundings in settings at different locations. CIT is again the agent of incidence, but the susceptible population is viewed as being the cells having development influence as modeled in Chapter 4. In this context, even settings with relatively little development can have their complexion altered substantially by presence of CIT as a major regional road. SaTScan output for the modified model is as follows.

```
SaTScan v9.1.1
```

```
Program run on: Sun Jul 17 09:27:38 2011

Purely Spatial analysis
scanning for clusters with high rates
using the Discrete Poisson model.
```

SUMMARY OF DATA

```
Study period.................: 2000/1/1 to 2000/12/31
Number of locations..........: 46
Total population.............: 1756433
Total number of cases........: 162281
Annual cases / 100000........: 9220.1
```

MOST LIKELY CLUSTER

```
1.Location IDs included.: 16, 15, 10, 22, 17, 9, 21, 11, 23,
                          14, 6, 28, 18, 20, 5, 27, 7
  Coordinates / radius..: (450265,4.58164e+006) / 11180.34
  Population............: 815114
  Number of cases.......: 120717
  Expected cases........: 75310.31
  Annual cases / 100000.: 14779.2
  Observed / expected...: 1.60
  Relative risk.........: 3.35
  Log likelihood ratio..: 26269.903924
  P-value...............: < 0.000000000000000010
```

SECONDARY CLUSTERS

```
2.Location IDs included.: 40
  Coordinates / radius..: (455265,4.60164e+006) / 0
  Population............: 42544
  Number of cases.......: 4932
  Expected cases........: 3930.74
  Annual cases / 100000.: 11568.7
  Observed / expected...: 1.25
  Relative risk.........: 1.26
  Log likelihood ratio..: 121.066064
  P-value...............: < 0.000000000000000010
```

PARAMETER SETTINGS

Input

```
  Case File          : D:\satscans\CITcases.cas
  Population File     : D:\satscans\CITabls.pop
  Coordinates File    : D:\satscans\Settings.geo
  Time Precision      : None
  Start Time          : 2000/1/1
  End Time            : 2000/12/31
  Coordinates         : Cartesian
```

Analysis

  Type of Analysis   : Purely Spatial
  Probability Model   : Discrete Poisson
  Scan for Areas with   : High Rates

Output

  Results File : D:\satscans\CIThi
  Cluster File : D:\satscans\CIThi.col.txt

Data Checking

  Temporal Data Check   : Check to ensure that all
cases and controls are within the specified temporal study
period.
  Geographical Data Check : Check to ensure that all
observations (cases, controls and populations) are within the
specified geographical area.

Spatial Neighbors

  Use Non-Euclidian Neighbors file : No
  Use Meta Locations File   : No
  Multiple Coordinates Type   : Allow only set of
coordinates per location ID.

Spatial Window

  Maximum Spatial Cluster Size : 50 percent of population at
risk
  Window Shape   : Circular
  Isotonic Scan   : No

Space And Time Adjustments

  Adjust for known relative risks : No

Inference

  P-Value Reporting   : Default Combination
  Adjusting for More Likely Clusters : No
  Number of Replications   : 999

Clusters Reported

  Criteria for Reporting Secondary Clusters : No Geographical
Overlap

```
Additional Output
-----------------
  Report Critical Values    : No
  Report Monte Carlo Rank   : No
  Print ASCII Column Headers : No

Run Options
-----------
  Processer Usage    : All Available Proccessors
  Logging Analysis   : Yes
  Suppress Warnings  : No
```

```
Program completed   : Sun Jul 17 09:27:40 2011
Total Running Time : 2 seconds
Processor Usage     : 2 processors
```

SaTScan has again found two significant clusters, but the primary cluster is quite different and consists of 17 settings. The secondary cluster is again setting 40 as a singleton. The map of clusters is shown in Figure 7.3.

FIGURE 7.3
Significant clusters (hotspots) for CIT contribution to complexion of development using Poisson model in SaTScan.

Figure 7.3 shows that all of the localities in the clusters are again present except for number 26. However, the primary cluster now includes several other localities to the south and east that were not included before. Several of the added localities such as 5, 6, and 9 are low spots rather than high spots according to the decile map of Figure 7.1. However, this second model considers only development-influenced cells in assigning relative risk. Localities 5, 6, and 9 are among those that owe much of their somewhat sparse development to local and regional roads that are included in the CIT category. Thus, it can be seen that choice of model is an important consideration.

7.5 Particular Proximity

For the binomial (Bernoulli) model, SaTScan allows specification of proximate points instead of the circular scanning strategy. This requires a file of neighbor numbers.

The Python© (2.7) Program 7.1 (see Section 7.7) takes a regular geographic file for SaTScan as input and produces an output file listing rook or queen neighbors for each point-by-point number. Rook neighbors are at the sides, and queen neighbors are at both sides and corners (as per chess moves). Entering grid spacing as a negative number gives queen neighbors.

Program 7.1 Python (2.7) program for rook or queen neighbors.

```
# Find rook or queen neighbors.
spacing = raw_input("Grid spacing?\n")
spacing = float(spacing)
corner = 0
nbrs = 0
if spacing < 0:
  corner = 1
  spacing *= -1.0
cntrfile = raw_input("Input file of centers?\n")
ptsfile = open(cntrfile,"r")
nabrfile = raw_input("Input file for neighbors?\n")
nbrfile = open(nabrfile,"w")
gridpts = []
for line in ptsfile:
  lining = line.split()
  ptnum = int(lining[0])
  ptx = float(lining[1])
  pty = float(lining[2])
  apt = (ptnum,ptx,pty)
  gridpts.append(apt)
  pts = len(gridpts)
```

```
feedbak = "Number of points" + " " + str(pts)
print(feedbak)
ptsfile.close
for item in gridpts:
  pnt = item[0]
  ptx = item[1]
  pty = item[2]
  sides = str(pnt)
  cornrs = ""
  cornr = 0
  for itm in gridpts:
    pntt = itm[0]
    ptxx = itm[1]
    ptyy = itm[2]
    if pntt != pnt:
      dsqr = (ptx - ptxx) * (ptx - ptxx)
      dsqr += (pty - ptyy) * (pty -ptyy)
      dst = pow(dsqr,0.5)
      if dst <= spacing:
        sides += " " + str(pntt)
      if dst > spacing and dst < 2.0*spacing:
        cornrs += " " + str(pntt)
        cornr = 1
  if corner == 1 and cornr == 1:
    sides += cornrs
  sides += "\n"
  nbrfile.write(sides)
nbrfile.close()
ending = raw_input("Press ENTER to finish")
```

The first few lines of the queen mode output file are as follows.

```
1 2 3
2 1 3 6 5 7
3 2 4 7 1 6 8
4 3 8 7
5 6 9 2 10
6 2 5 7 10 3 9 11
7 3 6 8 11 2 4 10 12
8 4 7 12 3 11
9 5 10 15 6 14 16
10 6 9 11 16 5 7 15 17
11 7 10 12 17 6 8 16 18
12 8 11 18 7 17
```

The binomial case and control files remain the same, but the neighbors file replaces the geographic points file. In this mode, SaTScan treats a point and its neighbors as a window. This more abstract kind of window can also be used in other connectivity contexts such as nodes on a network. SaTScan output corresponding to the previous CIT concentration run is as follows.

```
                    ┌─────────────────────────────┐
                    │        SaTScan v9.1.1        │
                    └─────────────────────────────┘

Program run on: Fri Aug 12 14:16:56 2011

Purely Spatial analysis
scanning for clusters with high rates
using the Bernoulli model.
```

SUMMARY OF DATA

```
Study period................: 2000/1/1 to 2000/12/31
Number of locations.........: 46
Total population............: 4600000
Total number of cases.......: 162281
```

MOST LIKELY CLUSTER

```
1.Location IDs included.: 21, 15, 20, 22, 27, 14, 16, 26, 28
   Population............: 900000
   Number of cases.......: 100815
   Expected cases........: 31750.63
   Observed / expected...: 3.18
   Relative risk.........: 6.74
   Log likelihood ratio..: 73674.305650
   P-value...............: < 0.000000000000000010
```

SECONDARY CLUSTERS

```
2.Location IDs included.: 7
   Population............: 100000
   Number of cases.......: 6774
   Expected cases........: 3527.85
   Observed / expected...: 1.92
   Relative risk.........: 1.96
   Log likelihood ratio..: 1263.098368
   P-value...............: < 0.000000000000000010

3.Location IDs included.: 40, 35
   Population............: 200000
   Number of cases.......: 9972
   Expected cases........: 7055.70
   Observed / expected...: 1.41
   Relative risk.........: 1.44
   Log likelihood ratio..: 584.190346
   P-value...............: < 0.000000000000000010
```

```
4.Location IDs included.: 38, 33
   Population............: 200000
   Number of cases.......: 7752
   Expected cases........: 7055.70
   Observed / expected...: 1.10
   Relative risk.........: 1.10
   Log likelihood ratio..: 36.159596
   P-value...............: 0.0000000000000013

5.Location IDs included.: 18
   Population............: 100000
   Number of cases.......: 3994
   Expected cases........: 3527.85
   Observed / expected...: 1.13
   Relative risk.........: 1.14
   Log likelihood ratio..: 31.362244
   P-value...............: 0.00000000000018
```

```
PARAMETER SETTINGS

Input
-----
   Case File        : D:\satscans\CITcases.cas
   Control File     : D:\satscans\CITctl.txt
   Time Precision   : None
   Start Time       : 2000/1/1
   End Time         : 2000/12/31

Analysis
--------
   Type of Analysis    : Purely Spatial
   Probability Model   : Bernoulli
   Scan for Areas with : High Rates

Output
------
   Results File : D:\satscans\CITbinabr
   Cluster File : D:\satscans\CITbinabr.col.txt

Data Checking
-------------
   Temporal Data Check     : Check to ensure that all
cases and controls are within the specified temporal study
period.
   Geographical Data Check : Check to ensure that all
observations (cases, controls and populations) are within the
specified geographical area.
```

```
Spatial Neighbors
------------------
  Use Non-Euclidian Neighbors file  : Yes
  Non-Euclidian Neighbors file      : D:\satscans\nabors.txt
  Use Meta Locations File           : Yes


Spatial Window
--------------
  Maximum Spatial Cluster Size  : 50 percent of population at
risk
    Isotonic Scan                   : No


Inference
---------
  P-Value Reporting                 : Default Combination
  Adjusting for More Likely Clusters : No
  Number of Replications            : 999


Clusters Reported
-----------------
  Criteria for Reporting Secondary Clusters : No Geographical
Overlap


Additional Output
-----------------
  Report Critical Values    : No
  Report Monte Carlo Rank   : No
  Print ASCII Column Headers : No


Run Options
-----------
  Processer Usage    : All Available Proccessors
  Logging Analysis   : Yes
  Suppress Warnings  : No
```

```
Program completed   : Fri Aug 12 14:19:25 2011
Total Running Time  : 2 minutes 29 seconds
Processor Usage     : 2 processors
```

SaTScan has not tried any window larger than a queen's square but has now produced several significant secondary clusters. Considering primary and secondary clusters, more places are now included in significant clusters, and they cover a much more irregular area as shown in Figure 7.4.

The most obvious anomalies in Figure 7.4 relative to the high spot map of deciles in Figure 7.1 are locality number 17 and locality number 29. These

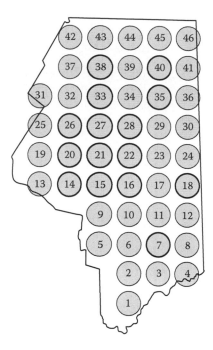

FIGURE 7.4

Significant clusters (hotspots) for concentration of CIT core development by the binomial (Bernoulli) model with neighbor windows in SaTScan.

are critically located at the corners of the significant queen block. Thus, it is apparent that the nature of the scanning window does make a difference.

7.6 Upper Level Set (ULS) Scanning

G. P. Patil and collaborating colleagues (Patil and Taillie 2004; Patil et al. 2009) have considered restrictive consequences of particular windows as seen above and developed adaptations of scanning to trees of upper level sets (ULS tree). The ULS tree approach effectively does successive slicing of the spatial topology along contours of a response surface. It formalizes the flexibility of scanning neighborhoods through extension to networks of neighbors of neighbors. In a spatial sense, this allows for scanning across narrow necks of connectivity, which can tend to give rise to somewhat stringy irregularly shaped clusters in much the manner as single linkage strategies of clustering. One aspect of their current research is continuing to allow for irregular clusters with some facility for favoring cluster compactness.

7.7 Backdrop: Python Programming

Due to its high-level structure and statistical orientation, **R** does not afford full control of processes and formats. This just affirms that no software system is ideal for everything. Python (www.python.org) is a more conventional programming environment without statistical predilection, which also has open-source public availability without cost, and thus complements **R** in several regards. The Python processor must be downloaded and installed on the host computer in order to run a Python program. The Python graphical user interface development environment is called IDLE. Python has a large user community that is quite active, and several books are available for learning Python (Ceder 2010; Dawson 2010; Lutz 2009; Lutz 2011) and for reference (Beazley 2009; Lutz 2010).

Python comes in several versions as a result of ongoing development, and the transition from 2.x to 3.x versions is nontrivial to the point of possibly requiring that some aspects of a program be rewritten. Python supports both object-oriented (OOP) and procedural approaches to programming. It is often used as a *scripting* language to bridge between and set up scenarios for other software systems. A spatially oriented development effort called Python Spatial Analysis Library (PySAL) has acquired some momentum. PySAL is an open source cross-platform library of spatial analysis functions written in Python. It is intended to support the development of high-level applications for spatial analysis. Doing an internet search for PySAL will serve to access appropriate websites. We use Python as an adjunct to **R** in situations that require relatively strict structure.

References

Beazley, D. *Python Essential Reference, 4th ed*. Old Tappan, NJ: Pearson Education, Inc., 2009.

Ceder, V. *The Quick Python Book, 2nd ed*. Greenwich, CT: Manning Publications, 2010.

Costa, M., and Kulldorff, M. Applications of spatial scan statistics: a review. Chapter 6. In: Glaz, J., Pozdnyakov, V., and Wallenstein, S., Eds. *Scan Statistics: Methods and Applications*. Boston: Birkhauser, 2009.

Dawson, M. *Python Programming, 3rd ed*. Boston: Course Technology, 2010.

Kulldorff, M. A spatial scan statistic. *Communications in Statistics: Theory and Methods* **26**: 1481–1496, 1997. [on-line]. Accessed August, 2011.

Lawson, A., and Denison, D., Eds. *Spatial Cluster Modelling*. Boca Raton, FL: Chapman & Hall/CRC, 2002.

Lutz, M. *Learning Python: Powerful Object-Oriented Programming.* Sebastopol, CA: O'Reilly Media, 2009.

Lutz, M. *Python Pocket Reference.* Sebastopol, CA: O'Reilly Media, 2010.

Lutz, M. *Programming Python.* Sebastopol, CA: O'Reilly Media, 2011.

Patil, G. P., Joshi, S., Myers, W., and Koli, R. ULS scan statistic for hotspot detection with continuous gamma response. Chapter 12. In: Glaz, J., Pozdnyakov, V., and Wallenstein, S., Eds. *Scan Statistics: Methods and Applications.* Boston: Birkhauser, 2009.

Patil, G. P., and Taillie, C. Upper level set scan statistic for detecting arbitrarily shaped hotspots. *Environmental and Ecological Statistics* **11**: 183–197, 2004.

Wallenstein, S. Joseph Naus: Father of the scan statistic. Chapter 1. In: Glaz, J., Pozdnyakov, V., and Wallenstein, S., Eds. *Scan Statistics: Methods and Applications.* Boston: Birkhauser, 2009.

8

Shape, Support, and Partial Polygons

8.1 Introduction

In Chapter 2, it was suggested that economies could be achieved in regard to size of computer files, utilization of memory resources, and simplified processing by using more parsimonious polygons that did not purport to approximate circles closely as a basis for integrated vicinity indicators (IVIs). It was also suggested that octagons would be a candidate to consider in this regard. A related concern is how changes in the area basis of integrative indicators giving them more local or broader scope are reflected in distributional changes for the values of the indicators. This latter is the *support* effect in the jargon of geostatistics and can be probed as a multiscale aspect of analyses (Longley and Batty 2003). Another aspect of area apportionment arises when the support structures span several partial polygons. These subjects are pursued in this chapter, which also provides a process review for compilation of IVIs.

8.2 Inscribed Octagons

Polygons inscribed in circles provide a practical framework for considering surrogate shapes for circles. A trigonometric analysis can be performed that relates each shape to its circumscribing circle in terms of the radius. The familiar area for a circle of radius r is the familiar πr^2. The corresponding area for an octagon is $2.82424r^2$, so the fractional area for the octagon is $2.82424/3.14159 = 0.89898$ or 89.898%, which rounds to 90%. Thus, 10% of a circular vicinity area around a point is lost in substituting an octagon. However, it was previously noted that ArcGIS used 315 points plus a repetition of the starting point for 316 points to represent the 2-km proximity perimeters used thus far. An octagon takes only eight points plus a repetition of the start for nine points. Thus, there is a saving in storage of $307/316 = 0.971519$ or more than 97% as the compensation for the 10% reduction in area.

As stated in Chapter 2, the geographic information system (GIS) does not provide internally for generating octagons as it does for circles. Therefore, an external facility must be constructed for generation in a manner that is amenable to importing into the GIS. Figure 8.1 shows an octagon with letters at the corners for reference in calculating coordinates. These computations are quite simple. Let X and Y be the coordinates of the center, with r as the radius of the circumscribing circle. Coordinates of the A, B, and C corners are as follows, with additions changed to subtractions as needed for the other corners.

$$X_A = X$$
$$Y_A = Y + r$$

$$X_B = X + 0.707107r$$
$$Y_B = Y + 0.707107r$$

$$X_C = X + r$$
$$Y_C = Y$$

ArcGIS does allow through its "Sample" tools for specifying polygons by corner coordinates computed externally, but the formatting is very particular and not well suited to the modalities of **R**. The Python (2.7) programming language is used here to generate octagons from an input file of centerpoint coordinates with output structured for import into ArcGIS. Program 8.1 gives Python code for this purpose. The first dozen lines of an appropriate textual input file of points are as follows.

```
1       450265.0626     4561637.415
2       450265.0626     4566637.415
3       455265.0626     4566637.415
4       460265.0626     4566637.415
```

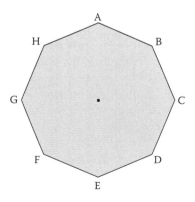

FIGURE 8.1
Octagon with lettered corners for reference in calculating coordinates.

5	445265.0626	4571637.415
6	450265.0626	4571637.415
7	455265.0626	4571637.415
8	460265.0626	4571637.415
9	445265.0626	4576637.415
10	450265.0626	4576637.415
11	455265.0626	4576637.415
12	460265.0626	4576637.415

Program 8.1 Python (2.7) program for generating octagons from center points.

```
# Generate octagons around center points
jj = 0
radius = raw_input("Radius of circumscribed circle?\n")
radus = float(radius)
cntrfile = raw_input("Input file of centers?\n")
ptsfile = open(cntrfile,"r")
octiv = raw_input("Output octagon file?\n")
octifile = open(octiv,"w")
omits = " 1.#QNAN 1.#QNAN"
octifile.write("Polygon\n")
for line in ptsfile:
    lining = line.split()
    print(lining)
    ptnum = int(lining[0])
    ptx = float(lining[1])
    pty = float(lining[2])
    ii = 0
    outing = str(jj) + " " + str(ii) + "\n"
    octifile.write(outing)
    xx = ptx
    yy = pty + radus
    outing = str(ii) + " " + str(xx) + " " + str(yy) + omits + "\n"
    octifile.write(outing)
    ii += 1
    xx = ptx + 0.707107 * radus
    yy = pty + 0.707107 * radus
    outing = str(ii) + " " + str(xx) + " " + str(yy) + omits + "\n"
    octifile.write(outing)
    ii += 1
    xx = ptx + radus
    yy = pty
    outing = str(ii) + " " + str(xx) + " " + str(yy) + omits + "\n"
    octifile.write(outing)
    ii += 1
    xx = ptx + 0.707107 * radus
    yy = pty - 0.707107 * radus
    outing = str(ii) + " " + str(xx) + " " + str(yy) + omits + "\n"
    octifile.write(outing)
    ii += 1
    xx = ptx
    yy = pty - radus
    outing = str(ii) + " " + str(xx) + " " + str(yy) + omits + "\n"
    octifile.write(outing)
    ii += 1
    xx = ptx - 0.707107 * radus
    yy = pty - 0.707107 * radus
    outing = str(ii) + " " + str(xx) + " " + str(yy) + omits + "\n"
```

```
    octifile.write(outing)
    ii += 1
    xx = ptx - radius
    yy = pty
    outing = str(ii) + " " + str(xx) + " " + str(yy) + omits + "\n"
    octifile.write(outing)
    ii += 1
    xx = ptx - 0.707107 * radius
    yy = pty + 0.707107 * radius
    outing = str(ii) + " " + str(xx) + " " + str(yy) + omits + "\n"
    octifile.write(outing)
    ii += 1
    xx = ptx
    yy = pty + radius
    outing = str(ii) + " " + str(xx) + " " + str(yy) + omits + "\n"
    octifile.write(outing)
    jj += 1
ptsfile.close()
octifile.write("END\n")
octifile.close()
ending = raw_input("Done -- press ENTER to close window.")
```

The output of the first two octagons having a radius of 2000 is as follows, wherein it is to be noted that numbering of the output polygons starts at 0 instead of 1 and the last two items on each line are placeholders for specifications that could be made but are not relevant to the current context. A line having only the word END follows the coordinate information for the last polygon.

```
Polygon
0  0
0  450265.0626  4563637.415  1.#QNAN  1.#QNAN
1  451679.2766  4563051.629  1.#QNAN  1.#QNAN
2  452265.0626  4561637.415  1.#QNAN  1.#QNAN
3  451679.2766  4560223.201  1.#QNAN  1.#QNAN
4  450265.0626  4559637.415  1.#QNAN  1.#QNAN
5  448850.8486  4560223.201  1.#QNAN  1.#QNAN
6  448265.0626  4561637.415  1.#QNAN  1.#QNAN
7  448850.8486  4563051.629  1.#QNAN  1.#QNAN
8  450265.0626  4563637.415  1.#QNAN  1.#QNAN
1  0
0  450265.0626  4568637.415  1.#QNAN  1.#QNAN
1  451679.2766  4568051.629  1.#QNAN  1.#QNAN
2  452265.0626  4566637.415  1.#QNAN  1.#QNAN
3  451679.2766  4565223.201  1.#QNAN  1.#QNAN
4  450265.0626  4564637.415  1.#QNAN  1.#QNAN
5  448850.8486  4565223.201  1.#QNAN  1.#QNAN
6  448265.0626  4566637.415  1.#QNAN  1.#QNAN
7  448850.8486  4568051.629  1.#QNAN  1.#QNAN
8  450265.0626  4568637.415  1.#QNAN  1.#QNAN
```

Toward an initial probing of the support effect, a second set of octagons is generated with a radius of 1500 m for the circumscribing circle. The initial 2-km circles have an area of 1257 ha, the 2-km octagons have 1130 ha, and

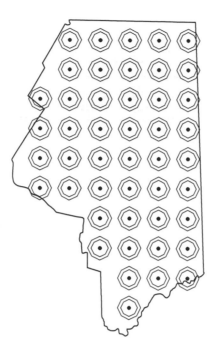

FIGURE 8.2
Octagons having a 2-km radius for the outer and 1.5-km radius for the inner.

635 ha for the 1.5-km octagons. Whereas the larger octagons have 90% of the circle area, the smaller octagons have only 50% of the circle area. The two sets of octagons are shown together in Figure 8.2.

8.3 Matching Margins and Adjusting Areas

Having imported the octagonal proximity perimeters into a GIS, IVIs cannot be composed until the margins of the buffers are matched to the boundary of the interest area by clipping and the areas of the buffers adjusted accordingly. In fact, simple importing of octagonal zones does not entail automatic calculation of the areas anyhow.

As for circular zones in Chapter 2, matching of the margins is accomplished by using the county boundary to clip the octagons via the "Analysis" tools in the GIS. Areas are then calculated using the "Calculate Geometry" capability of the GIS attribute table.

The clipped 2000-m octagons are shown in Figure 8.3, and the corresponding clipped 1500-m octagons are shown in Figure 8.4. Whereas several 2000-m zones have been clipped, only two of the 1500-m buffers have been noticeably clipped

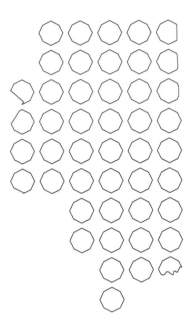

FIGURE 8.3
Clipped 2000-m (radius) octagonal proximity perimeters.

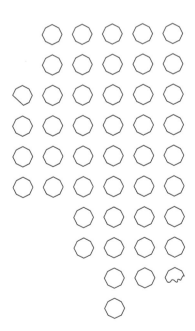

FIGURE 8.4
Clipped 1500-m (radius) octagonal proximity perimeters.

with another being slightly clipped. At this juncture, the new versions of proximity perimeters are ready for compiling IVIs with respective levels of support.

8.4 Shape and Support for Local Roads

We proceed to conduct a small investigation regarding shape and support as it relates to local roads. This context is chosen in part because it is both locally variable and suitable for visual presentation as well as numerical/statistical analysis. Figure 8.5 shows the (clipped) octagons of 2000-m radius mapped on the local roads that occur in the (clipped) 2000-m-radius circular zones. Likewise, Figure 8.6 shows the (clipped) octagons of 1500-m radius mapped on local roads that occur in the (clipped) 2000-m-radius circular zones. It is visually evident that omissions from the 2000-m-radius octagons are minor, whereas differences are more obvious for smaller octagons.

The Pearson correlation matrix for local roads with different shapes and sizes is given in Table 8.1, with LrdC2000 being the local road density IVI for circles of 2000-m radius, LrdO2000 being the local road density IVI for octagons of 2000-m radius, and LrdO1500 being the local road density IVI for octagons of 1500-m radius.

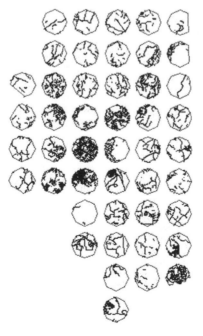

FIGURE 8.5
Clipped 2000-m octagons on local roads in (clipped) 2000-m circles.

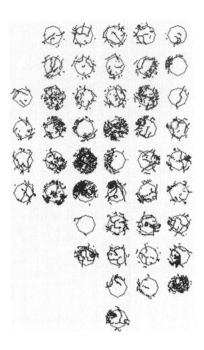

FIGURE 8.6
Clipped 1500-m octagons on local roads in (clipped) 2000-m circles.

TABLE 8.1

Pearson Correlation Coefficients for Local Road Densities
of Different Support

```
> cor(ShpSizLrds[,4:6],method="pearson")
          LrdC2000   LrdO2000   LrdO1500
LrdC2000  1.0000000  0.999006   0.9736628
LrdO2000  0.9990059  1.000000   0.9802081
LrdO1500  0.9736628  0.980208   1.0000000
```

The corresponding Spearman (rank) correlation matrix is in Table 8.2, and paired plots are given in Figure 8.7.

There is little difference evident for the 2000-m (radius) circle and octagon having the same radius covering 90% of the area. There is very strong correlation, but there are more evident differences between the 2000-m radius and the octagon of 1500-m radius covering roughly 50% of the larger circular area.

As might be expected, there is little loss of contextual information in approximating a circle by an octagon, which takes less than 3% of the coordinate storage capacity. If there is concern for the small loss of support, the radius can be increased by 1.054 to compensate. The effect of reducing the support area by a factor of 2 will depend on the spatial auto-association of the phenomenon under consideration.

TABLE 8.2

Spearman Correlation Coefficients for Local Road Densities
of Different Support

```
> cor(ShpSizLrds[,4:6],method="spearman")
            LrdC2000    LrdO2000    LrdO1500
LrdC2000  1.0000000  0.9955597  0.9261178
LrdO2000  0.9955597  1.0000000  0.9460993
LrdO1500  0.9261178  0.9460993  1.0000000
```

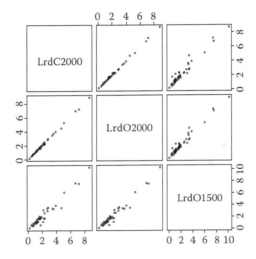

FIGURE 8.7
Panel of scatterplots for pairs of local road density indicators having different combination of
shape and support. LrdC2000 is for the 2000-m circle, LrdO2000 is for the 2000-m octagon, and
LrdO1500 is for the 1500-m (radius) octagon.

8.5 Precedence Plot for Shapes and Supports

Precedence properties provide one avenue for exploring effects of shape and
support in a multiscale manner. The partial precedence methodology introduced in Chapter 3 is also applicable for this purpose.

The first step toward obtaining a partial precedence plot is to apply the
POprecdn() function.

```
> PlacIDs <- ShpSizLrds[,1]
> ShpSizPrcdn <- POprecdn(PlacIDs,ShpSizLrds[,4:6],BigBetr)
```

A precedence plot is then obtained via the PrecPlot() function as depicted
in Figure 8.8 with labeling as may be appropriate.

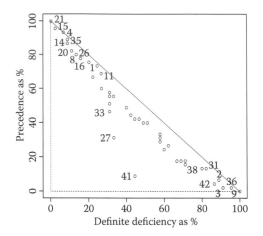

FIGURE 8.8
Precedence plot for shape and support of local road density indicators.

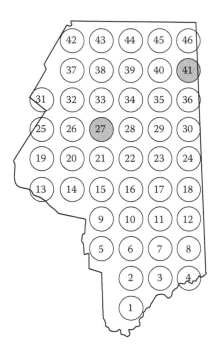

FIGURE 8.9
Sensitive settings for shape and support.

```
> PrecPlot(ShpSizPrcdn)
> identify(ShpSizPrcdn[,3],ShpSizPrcdn[,2])
```

Since the line numbers match the ID numbers for settings, the labels in Figure 8.8 show which settings have high precedence, low precedence, and particular sensitivity to shape and support. The greatest sensitivity is shown by settings 27 and 41, which are highlighted in Figure 8.9. Referring back to Figure 8.6, these can be seen to have the local roads concentrated in the periphery of the zones.

Such support effects are important to geostatistics (Isaaks and Srivastava 1989; Webster and Oliver 1990; Webster and Oliver 2001), which are given consideration in later chapters.

8.6 Supports Spanning Several Partial Polygons

Clipping of support structures to lie within an area of interest has been our standard strategy in order to avoid edge bias. Related reasoning applies when support structures such as circles or octagons span several partial polygons serving as sources of indicator information.

The spanning situation often occurs when compiling components of human habitation where statistical data pertain to minor civil divisions or census blocks and tracts. It also arises for environmental effects associated with watersheds and terrain types. Figure 8.10 shows 2-km-radius octagonal

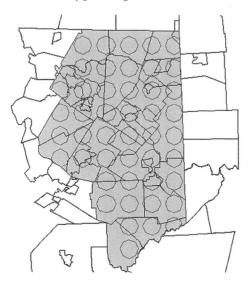

FIGURE 8.10
Octagonal support structures with 2-km radius in relation to municipal boundaries for Lackawanna County showing that supports span several partial polygons.

support structures superimposed on municipal boundaries in Lackawanna County. Settings 27 and 33, for example, each span four different municipal divisions.

The appropriate approach is akin to that for linear features described in detail earlier with some adaptations. The attributes of interest for each polygon should first be expressed on an area basis, such as population per square mile or square kilometer. The support structures are then intersected with the polygons, so that each piece of a support structure as a partial polygon is separately identified as a spatial segment. The area of each of the spatial segments is then determined and used to calculate an area-weighted contribution of that segment to the support structure. These contributions are then summarized over segments for each support structure.

References

Isaaks, E. H., and Srivastava, R. M. *An Introduction to Applied Geostatistics*. New York: Oxford Univ. Press, 1989.

Longley, P., and Batty, M., Eds. *Advanced Spatial Analysis: The CASA Book of GIS*. Redlands, CA: ESRI Press, 2003.

Webster, R., and Oliver, M. *Statistical Methods in Soil and Land Resource Survey*. New York: Oxford Univ. Press, 1990.

Webster, R., and Oliver, M. *Geostatistics for Environmental Scientists*. New York: Wiley, 2001.

9

Semisynchronous Signals
and Variant Vicinities

9.1 Introduction

In opening this chapter, we advocate for an alternative virtual view of simulated sentinels that send signals from their stations regarding status of the setting in which they are situated. The posting point localities are the sentinel stations, and the vicinity variates are the signals. This conceptual context is also conducive to configuring several series of sentinels that might, for example, be strategically stationed along crucial corridors. The precedence paradigm can provide interpretive interfaces for sentinel systems.

The overall suite of signals may have some with strong synchrony across substantial sets of stations, whereas others may be only semisynchronous. There may be a few particularly variant vicinities or some special sectors. A pervasive pattern of discordance among signals, however, would require segregation of signals into subsets. When there is stronger synchrony within some subsets of signals, this can be considered as *associative asynchrony* for the subsets. If an individual signal shows little or no synchrony with any other, then this would constitute an *idiosyncratic indicator*. Variant vicinities exhibit *anomalous asynchrony*. A *particular perspective* arises when there is synchrony for some subset of signals in some subset of station settings. Such a particular perspective will be evidenced by strong rank correlation for these signals when computed over the subset of stations.

We have shown in Chapter 3 how top and toe of precedence plots can be used to prioritize localities (sentinel stations). That was extended in Chapter 5 for groupwise (cluster) prioritization purposes. Chapter 8 has subsequently showed how diagonal displacement in precedence plots can serve to show sensitivity to differences in signals (as affected by spatial support). We continue by combining prioritization and sensitivity as an interpretive interface for sentinel signaling systems, for which we use five signals drawn from foregoing chapters. We take regional and local road densities (RrdIVI and LrdIVI) from Chapter 2, development density and modeled development influence (DvlpIVI and DvlmIVI) from Chapter 4, and intensity of low-density development (LoddIVI) from Chapter 6 as mean frequency of low-density residential development in a 450-m radius (15 cells) from a focal cell. Inasmuch as this latter indicator has not yet been mapped as deciles on sentinel stations, this is shown in Figure 9.1.

The leading lines for a data frame of ranks are given in Table 9.1.

A precedence plot for these five signals is shown in Figure 9.2, with selected stations identified. Neighboring stations 21, 15, and 14 are at the top with respect to composite human influence. Scattered stations 9, 2, 36, and 42 are in the toe. Stations 4, 7, and 38 have extreme diagonal displacements. Station 16 is also situated at the extreme left for its level of paired precedence. Strong

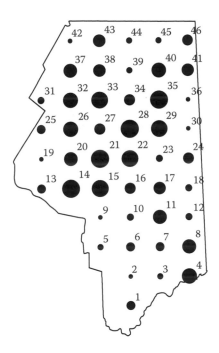

FIGURE 9.1
Deciles of LoddIVI indicator for mean frequency of low-density residential development in a 450-m radius.

TABLE 9.1

Leading Lines of Data Frame Having Five Vicinity Variates

	RrdIVI	LrdIVI	DvlpIVI	DvlmIVI	LoddIVI
1	3.0	35	16	13	20.0
2	1.0	6	3	4	3.5
3	27.0	5	19	12	7.0
4	2.0	43	34	43	36.0
5	4.5	31	4	2	8.0
6	20.0	16	17	11	18.0

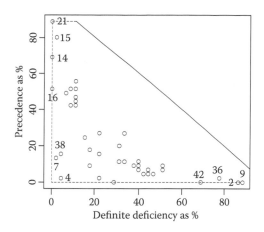

FIGURE 9.2
Precedence plot for five signals RrdIVI, LrdIVI, DvlpIVI, DvlmIVI, and LoddIVI with selected stations identified.

diagonal displacements reflect variant vicinities having anomalous combinations of the indicator signals.

There is a strong empty crescent to the left of the diagonal limiting line, which reflects substantial asynchrony among the indicators. The extent to which aggregate asynchrony is attributable to any particular signal can be expressed in terms of indefinite instances. The total of indefinite instances for all stations is 2155.57 with five signals, which is 46.86% of what is possible. Consider the precedence plot after deleting the LoddIVI signal as shown in Figure 9.3. This deletion changes the precedence plot relatively little except for elevating station 22 to the same level as station 14. The total of

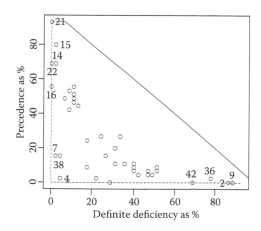

FIGURE 9.3
Precedence plot of four signals RrdIVI, LrdIVI, DvlpIVI, and DvlmIVI with selected stations identified.

indefinite instances has become 2497.78 with four signals, which is 54.29% of what is possible. Thus, the relative indefiniteness has increased rather than decreased as a result of deleting the LoddIVI signal, and this is evidently not an idiosyncratic indicator.

In the absence of a priori questions about specific signals, some suggestion about doing deletion may be sought in the rank correlation matrix. The rank correlation matrix for the current context appears in Table 9.2.

Because LoddIVI has high rank correlation with two other signals, deleting it should not be expected to have much impact on the overall precedence picture. Deleting the RrdIVI signal might be more interesting to explore. Accordingly, Figure 9.4 shows the precedence plot for the last four signals.

The precedence picture has now changed more substantially. Stations 28 and 35 have moved up to join 14, 15, and 21, but the same four are still at the toe. Stronger synchrony is evident by the lower-left "corner of confusion" being empty. The total of indefinite instances has dropped to 1368.89, which is 29.76% of what is possible.

TABLE 9.2

Rank Correlations for Five Vicinity Variates

RankCor	RrdIVI	LrdIVI	DvlpIVI	DvlmIVI	LoddIVI
RrdIVI	1.0000000	0.3137315	0.6630485	0.6608899	0.5564305
LrdIVI	0.3137315	1.0000000	0.7124884	0.6844897	0.7584508
DvlpIVI	0.6630485	0.7124884	1.0000000	0.9552266	0.9565679
DvlmIVI	0.6608899	0.6844897	0.9552266	1.0000000	0.9430473
LoddIVI	0.5564305	0.7584508	0.9565679	0.9430473	1.0000000

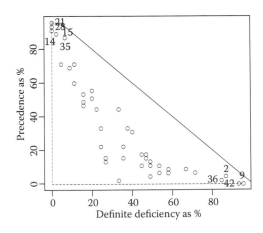

FIGURE 9.4

Precedence plot of four signals LrdIVI, DvlpIVI, DvlmIVI, and LoddIVI with selected stations identified.

Having worked with precedence plots as an interpretive interface for sentinel signaling makes other innovative approaches to synchrony of signals also of interest. Myers and Patil (2010) put forward several such strategies.

9.2 Distal Data

Determination of distal data starts with a table (data frame) of rank data for the several signals. The process proceeds by locating the two largest and two smallest ranks for each row (sentinel setting). These are the distal (most noncentral) data for that row. Let us call the largest rank as the *major maximal* rank and denote it by MM. Let the second largest rank be called the *minor maximal* rank and denoted by Mm. Similarly, let the smallest rank be called the *major minimal* rank and denoted by mM, with the second smallest being the *minor minimal* and denoted by mm. The logic of the approach is that major and minor being close together at either extremity implies absence of asynchrony.

Next, determine *upper extremity* as

$$M\Delta = MM - Mm$$

and the lower extremity as

$$m\Delta = mM - mm$$

with the upper extremity being positive or zero and the lower extremity being negative or zero. An extremity is zero when there are ties for major and minor. A large magnitude for $M\Delta$ or $m\Delta$ is indicative of asynchrony.

Accordingly, a *distal dataset* is to be constructed. Determine $M\Delta$ and $m\Delta$ for each row (sentinel) and identify the major signal sources (vicinity variates). Assign $M\Delta$ to the major maximal source and $m\Delta$ to the major minimal source. Assign a value of zero to all nonmajor sources. The **R** Function 9.1 named DstlData serves to compute a distal dataset from a data frame of ranks containing only the columns of ranks.

Function 9.1 DstlData function for compiling a data frame of distal data.

```
DstlData <- function(RankData)
# RankData is a data frame of ranks.
{Rnkcols <- length(RankData)
 Cases <- length(RankData[,1])
 XtrmXtnt <- RankData
 for(I in 1:Cases)
  {CaseRnks <- rep(0,Rnkcols)
   for(J in 1:Rnkcols)
    CaseRnks[J] <- RankData[I,J]
```

```
MinRnk <- min(CaseRnks)
MaxRnk <- max(CaseRnks)
Maxx <- 0
Minn <- 0
for(J in 1:Rnkcols)
  {if(CaseRnks[J]>=MaxRnk) Maxx <- J
   if(CaseRnks[J]<=MinRnk) Minn <- J
  }
CaseRnks <- sort(CaseRnks)
LoEE <- CaseRnks[1] - CaseRnks[2]
HiEE <- CaseRnks[Rnkcols] - CaseRnks[Rnkcols-1]
for(J in 1:Rnkcols)
  XtrmXtnt[I,J] <- 0
if(LoEE<0) XtrmXtnt[I,Minn] <- LoEE
if(HiEE>0) XtrmXtnt[I,Maxx] <- HiEE
}
XtrmXtnt
}
```

Parallel boxplots of the distal data are shown in Figure 9.5. The RrdIVI regional roads indicator shows three outliers on the low side but none on the high side, which suggests the presence of some anomalous asynchrony. The sentinels responsible for the anomalies can be determined by making an index plot of the distal data on this indicator as shown in Figure 9.6. Sentinel number 4 is one of the apparent anomalies, with the other being number 8. The third outlier shown in the boxplot of Figure 9.5 is not particularly prominent. The rank rows for the two anomalies are as follows, both of which have low ranks for RrdIVI with considerably higher ranks for other signals.

RrdIVI	LrdIVI	DvlpIVI	DvlmIVI	LoddIVI
4	2	43	34	43	36
8	6	36	31	27	33

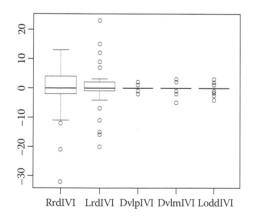

FIGURE 9.5
Parallel boxplots for distal data on five signals.

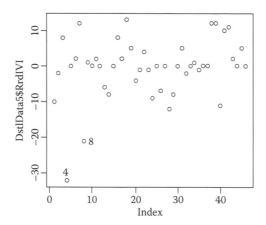

FIGURE 9.6
Index plot for distal data on RrdIVI indicator with anomalies identified.

The larger interquartile box for RrdIVI in Figure 9.5 also shows that this signal exhibits a propensity for larger distal differences and, therefore, less synchrony than the other signals.

The distal data do not offer strong evidence of anomalies for the last three indicators. For local roads, however, there is more involved than anomalies because more than 20% of settings show as outliers. This suggests inspecting a scatterplot of the roads ranks as shown in Figure 9.7 to see if there are interesting interactions.

Figure 9.7 shows that, in addition to sentinels 4 and 8, sentinels 1 and 5 also have higher ranks for density of local roads than regional roads. Conversely, sentinels 7 and 38 have higher ranks for densities of regional roads than local

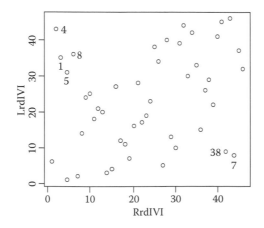

FIGURE 9.7
Plot of local road density ranks on *Y* versus regional road density ranks on *X*.

TABLE 9.3

Rank Correlations after Six Stations Are Segregated

	RrdIVI	LrdIVI	DvlpIVI	DvlmIVI	LoddIVI
RrdIVI	1.0000000	0.6601395	0.7608550	0.7745885	0.6915274
LrdIVI	0.6601395	1.0000000	0.8211887	0.8073534	0.8254827
DvlpIVI	0.7608550	0.8211887	1.0000000	0.9581331	0.9638292
DvlmIVI	0.7745885	0.8073534	0.9581331	1.0000000	0.9604810
LoddIVI	0.6915274	0.8254827	0.9638292	0.9604810	1.0000000

roads. When these six sentinels are segregated, the rank correlation matrix is modified (Table 9.3).

The pronounced lack of synchrony between the two road density signals has been removed from the modified matrix. Thus, it can be said that there is anomalous asynchrony due to interaction of road signals for these six situations, giving reason to segregate them. The last three signals constitute a particular perspective.

9.3 Median Models

Median models of signal structure are a little like cross-validation in the respect of predicting a signal for a sentinel as the median rank of other signals for that sentinel. Interest then lies in the *synchronization shift* that is needed to move from the actual rank to the modeled rank.

Synchronization shift = Modeled rank – Actual rank

Given rank data as input, the **R** Function 9.2 named MdnMdlSS implements the median model and computes the synchronization shifts. Correlations and scatterplots for the sync shifts can provide insights that cannot be easily obtained directly from the actual ranks. As a preliminary to that pursuit, we recall that the foregoing has shown positive rank correlations of varying degree among all of the five signals.

Function 9.2 MdnMdlSS function for synchronization shifts of median models.

```
MdnMdlSS <- function(RankData)
# Computes median model signal shifts.
{Rnkcols <- length(RankData)
 Cases <- length(RankData[,1])
 RnkData <- RankData
 for(I in 1:Cases)
```

```
{CaseRnks <- rep(0,Rnkcols)
 for(J in 1:Rnkcols)
   CaseRnks[J] <- RankData[I,J]
 for(J in 1:Rnkcols)
  {Medn <- median(CaseRnks[-J])
   SyncShft <- Medn - CaseRnks[J]
   RnkData[I,J] <- SyncShft
  }
 }
RnkData
}
```

A lattice of paired plots for the synchronization shifts is shown in Figure 9.8, and a correlation matrix for synchronization shifts is given in Table 9.4.

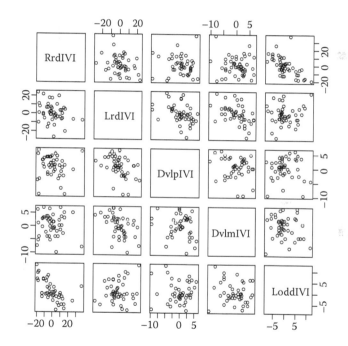

FIGURE 9.8

Paired plots for synchronization shifts of median models.

TABLE 9.4

Correlation Matrix for Synchronization Shifts

	RrdIVI	LrdIVI	DvlpIVI	DvlmIVI	LoddIVI
RrdIVI	1.00000000	-0.2661920	-0.066869074	-0.04250767	-0.614254465
LrdIVI	-0.26619205	1.0000000	-0.386372998	-0.48558635	0.108763493
DvlpIVI	-0.06686907	-0.3863730	1.000000000	-0.11804568	-0.008488451
DvlmIVI	-0.04250767	-0.4855863	-0.118045679	1.00000000	-0.158115554
LoddIVI	-0.61425447	0.1087635	-0.008488451	-0.15811555	1.000000000

Contrary to the parent correlations, most of the correlations for synchronization shifts are negative. Thus, the shifts are reflecting second-order effects that were not apparent in the rank signals. We can consider the rank signals as expressing "overtones" and the synchronization shifts as expressing "undertones."

The strongest of these undertones is contrasting the RrdIVI regional roads density with the LoddIVI density of low-density residential development. A possible causality for this second-order effect is that there may be some preference not to have low-density residential developments situated in close proximity to major roads so as to avoid heavier traffic.

There are also undertones contrasting local road density (LrdIVI) with regional road density, and contrasting overall development (DvlpIVI) and modeled development influence (DvlmIVI). It is again not difficult to find some speculative causality for these. In the absence of regional roads, transportation relies on local roads. Many local roads are subsumed in commercial/industrial/transportation, but local roads also occur in the absence of development influence. Correlative undertones are not notable for the particular perspective of the last three indicators.

There are only two obvious anomalies for the RrdIVI and LoddIVI undertones, and it is inviting to examine these more specifically. A plot of these two shifts is shown in Figure 9.9, with the anomalies identified. Not surprisingly, the anomalies again prove to be settings 4 and 8, with 4 being particularly prominent.

When station 4 is segregated, the correlation matrix shows this undertone becoming quite strong as follows, which reflects the somewhat anomalous nature of RrdIVI relative to the others as observed earlier by changes in precedence plots when RrdIVI is deleted (Table 9.5).

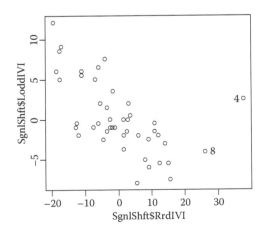

FIGURE 9.9
Scatterplot of synchronization shifts for RrdIVI and LoddIVI signals.

TABLE 9.5

Correlation for Synchronization Shifts without Station 4

	RrdIVI	LrdIVI	DvlpIVI	DvlmIVI	LoddIVI
RrdIVI	1.0000000	-0.2397541	-0.20288077	0.12494116	-0.74455704
LrdIVI	-0.2397541	1.0000000	-0.37170130	-0.55245491	0.11919894
DvlpIVI	-0.2028808	-0.3717013	1.00000000	-0.05386309	-0.02639114
DvlmIVI	0.1249412	-0.5524549	-0.05386309	1.00000000	-0.14168643
LoddIVI	-0.7445570	0.1191989	-0.02639114	-0.14168643	1.00000000

Investigation of synchronization shifts should not be keyed entirely on correlative effects. The paired plot of shifts for RrdIVI against DvlmIVI does not show correlation, but some stations are set apart, as labeled in Figure 9.10. Station number 4 is seen to be the one that is set apart in a far corner, and station number 8 is somewhat set apart to the right. Additionally, station 37 is set apart at the bottom.

Even though DvlpIVI and DvlmIVI show strong synchrony, there are still three stations that are set strongly apart in an opposite corner. Figure 9.11 has these three stations labeled.

The rank data for stations labeled in Figure 9.11 are as follows.

	RrdIVI	LrdIVI	DvlpIVI	DvlmIVI	LoddIVI
3	27	5	19	12	7
20	25	38	39	29	34
40	16	27	37	28	35

Station 3 has lower ranks for LrdIVI and LoddIVI, but any patterns for stations 20 and 40 are seemingly subtle.

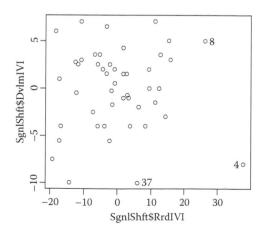

FIGURE 9.10
Scatterplot of synchronization shifts for RrdIVI and DvlmIVI signals.

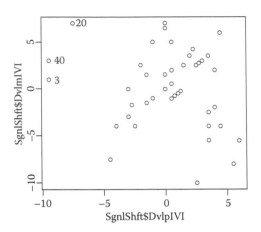

FIGURE 9.11
Scatterplot of synchronization shifts for DvlpIVI and DvlmIVI signals.

9.4 Pairing/Placement Patterns of Signal Strengths

We now pursue patterns of signal synchrony in a combinatorial context by multivariate methods based on ranking ranks (cross-ranking). We return to the rank data of the five signals obtained by ranking the data in each column, for which the head of the file as given earlier is as follows. The columns containing decimals are ones in which ties have occurred in the ranking.

	RrdIVI	LrdIVI	DvlpIVI	DvmIVI	LoddIVI
1	3.0	35	16	13	20.0
2	1.0	6	3	4	3.5
3	27.0	5	19	12	7.0
4	2.0	43	34	43	36.0
5	4.5	31	4	2	8.0
6	20.0	16	17	11	18.0

The first step is to rank the ranks in each row (sentinel) without reference to the ranks in any other row. Function 9.3 named CrosRank does this in **R**.

Function 9.3 R CrosRank function facility.

```
CrosRank <- function(RankData)
# RankData is a data frame of ranks.
# Output is cross-ranking by case.
{Rnkcols <- length(RankData)
 Cases <- length(RankData[,1])
```

```
for(I in 1:Cases)
 {CaseRnks <- rep(0,Rnkcols)
  for(J in 1:Rnkcols)
    CaseRnks[J] <- RankData[I,J]
  CaseRnks <- rank(CaseRnks)
  for(J in 1:Rnkcols)
    RankData[I,J] <- CaseRnks[J]
 }
 RankData
}
```

The head of the output file obtained by applying the CrosRank function is as follows.

	RrdIVI	LrdIVI	DvlpIVI	DvlmIVI	LoddIVI
1	1	5.0	3	2.0	4
2	1	5.0	2	4.0	3
3	5	1.0	4	3.0	2
4	1	4.5	2	4.5	3
5	3	5.0	2	1.0	4
6	5	2.0	3	1.0	4

Each row and column now contains only rank numbers less than or equal to five. In the head of the file shown immediately above, RrdIVI has the lowest rank three times and the highest rank twice, with only the third line (sentinel) having an intermediate rank. If such a propensity toward particular positions is pervasive in the dataset, then other signals will necessarily also show pattern of placement to some degree. The present pattern of peripheral placement would imply some asynchrony for RrdIVI relative to other signals. Here we seek to acquire insights from any patterns of pairing among signals as well as patterns of particular placement.

The strategy proceeds by indexing signals in occurrence order of cross-ranks for each row (sentinel) without trying to untangle ties. Ties can partially perturb placements, but do not affect pairings of placement among the signals. Function 9.4 for **R** named CrosNdx gives an index file as output for visual examination and explicit exploration, but the indexing is incorporated in subsequent steps as a passing process while working directly from the cross-ranks as input. We apply it here to help lend transparency to the procedure.

Function 9.4 R CrosNdx function facility for indexing signals.

```
CrosNdx <- function(CrosRnks)
# CrosRnks is a data frame of cross ranks.
# Output is CrosIndx data with row-wise signal numbers.
{Rnkcols <- length(CrosRnks)
 Cases <- length(CrosRnks[,1])
 for(I in 1:Cases)
   {CaseRnks <- rep(0,Rnkcols)
    for(J in 1:Rnkcols)
      CaseRnks[J] <- CrosRnks[I,J]
```

```
    CasOrdr <- order(CaseRnks)
    for(J in 1:Rnkcols)
      CrosRnks[I,J] <- CasOrdr[J]
    }
  CrosNams <- paste("Indx",1:Rnkcols,sep="")
  names(CrosRnks) <- CrosNams
  CrosRnks
  }
```

Function 9.5 for **R** named CrosCnts makes a tabulation of how often each signal is indexed in each position among all instances (sentinels or stations).

Function 9.5 R CrosCnts function facility for frequency of placement.

```
CrosCnts <- function(CrosRnks)
# CrosRnks is a data frame of cross ranks.
# Output is Crosplac matrix with rows as signals & columns as ranks.
{Rnkcols <- length(CrosRnks)
 Cases <- length(CrosRnks[,1])
 Elemnts <- Rnkcols * Rnkcols
 Crosplac <- rep(0,Elemnts)
 Crosplac <- matrix(data=Crosplac,nrow=Rnkcols,ncol=Rnkcols)
 for(I in 1:Cases)
   {CaseRnks <- rep(0,Rnkcols)
    for(J in 1:Rnkcols)
      CaseRnks[J] <- CrosRnks[I,J]
    CasOrdr <- order(CaseRnks)
    for(J in 1:Rnkcols)
     {K <- CasOrdr[J]
      Crosplac[K,J] <- Crosplac[K,J] + 1
      }
   }
 Crosplac
 }
```

Applying the CrossNdx function gives the following as head of the output file.

	Indx1	Indx2	Indx3	Indx4	Indx5
1	1	4	3	5	2
2	1	3	5	4	2
3	2	5	4	3	1
4	1	3	5	2	4
5	4	3	1	5	2
6	4	2	3	5	1

The above indexing shows that signal number 1 (RrdIVI) has the lowest order of placement in sentinels 1 and 2, whereas signal number 2 (LrdIVI) has the highest order of placement. These placements are reversed for sentinel 3. Thus, the road signals occupy opposite extremes in the first three sentinels. A complete count of placements as obtained from the CrosCnts function is as follows.

S\P	[,1]	[,2]	[,3]	[,4]	[,5]
[1,]	17	3	3	5	18
[2,]	13	8	4	9	12

[3,]	6	13	19	5	3
[4,]	6	13	9	13	5
[5,]	4	9	11	14	8

The above tabulation has signals as its rows and placement positions (of cross-ranks) as its columns. Thus, the first signal (RrdIVI) is at the extremes 35 of 46 times, indicating its apparent asynchrony. The second signal (LrdIVI) is at the extremes 25 of 46 times, which is less than RrdIVI but still more than half. The particular perspective of three development indicators noted earlier has a propensity for middle placement.

The CrosCpld function facility presented as Function 9.6 serves as the primary pattern prospector for pairing. The first input consists of the cross-ranks, and the second input is a pattern probe having as many elements as there are signals.

Function 9.6 R CrosCpld function facility for patterns of pairing.

```
CrosCpld <- function(CrosRnks,Placing)
# CrosRnks is a data frame of cross ranks.
# Placing is a pattern probe vector like c(1,1,1,1) for 4 signals.
# Output is Croscupl matrix with rows as signals & columns as places.
{Rnkcols <- length(CrosRnks)
 Cases <- length(CrosRnks[,1])
 Elemnts <- Rnkcols * Rnkcols
 Croscupl <- rep(0,Elemnts)
 Croscupl <- matrix(data=Croscupl,nrow=Rnkcols,ncol=Rnkcols)
 for(I in 1:Cases)
   {CaseRnks <- rep(0,Rnkcols)
    for(J in 1:Rnkcols)
     CaseRnks[J] <- CrosRnks[I,J]
    CasOrdr <- order(CaseRnks)
    for(J in 1:Rnkcols)
     {if(Placing[J]>0)
       {K <- CasOrdr[J]
        JJ <- J-1
        if(JJ>0 && Placing[JJ]>=0)
         {L <- CasOrdr[JJ]
          Croscupl[L,K] <- Croscupl[L,K] + 1
         }
        JJ <- J+1
        if(JJ<=Rnkcols && Placing[JJ]>=0)
         {L <- CasOrdr[JJ]
          Croscupl[L,K] <- Croscupl[L,K] + 1
         }
       }
     }
    L <- CasOrdr[1]
    if(Placing[1]>0) Croscupl[L,L] <- Croscupl[L,L] - 1
    L <- CasOrdr[Rnkcols]
    if(Placing[Rnkcols]>0) Croscupl[L,L] <- Croscupl[L,L] - 1
   }
 Croscupl
}
```

Using a pattern probe having ones in all elements will give an output matrix containing frequencies for all pairs of signals as follows. The diagonal would have zeros because a signal can only be in one position and therefore cannot pair with itself. This is exploited to show as negative numbers the frequency with which a signal appears in end positions and thus has only one pairing partner instead of two.

```
> PatrnProbe <- c(1,1,1,1,1)
> CrosCpld(CrosRank5,PatrnProbe)
      [,1] [,2] [,3] [,4] [,5]
[1,]   -35   17    9   18   13
[2,]    17  -25   19   11   20
[3,]     9   19   -9   30   25
[4,]    18   11   30  -11   22
[5,]    13   20   25   22  -12
```

As seen earlier, RrdIVI occurs at the extremes in 35 of 46 settings. The only apparent pattern in pairing for RrdIVI is lower pairing with the development signal (DvlpIVI). Presumably, this reflects the modern preference for regional roads to bypass heavily urbanized areas. The last three signals comprising the particular perspective all have high pairing propensities with each other. The local roads signal (LrdIVI) does not exhibit much in the way of preferential pairing.

The pattern probe can be made more specific through the concept of *pivot* positions and *pairing* positions. Pivot position is the reference placement, and pairing positions are the placement positions before or after the pivot. If the first position (placement) is the pivot, then the only pairing position is immediately after. If the last position is the pivot, then the only pairing position is immediately before. In all other pivot positions there are pairing positions before and after. If a position has a positive number in the pattern probe, then it has eligibility as both pivot and pairing. If a position has a zero in the pattern probe, then has eligibility only for pairing. If a position has a negative number, then it is excluded for both pivot and pairing. To look only at the three central positions, we can use c(-1,1,1,1,-1) as a probe.

```
> PatrnProbe <- c(-1,1,1,1,-1)
> CrosCpld(CrosRank5,PatrnProbe)
      [,1] [,2] [,3] [,4] [,5]
[1,]     0    1    3    4    6
[2,]     1    0   12    5    7
[3,]     3   12    0   22   19
[4,]     4    5   22    0   13
[5,]     6    7   19   13    0
```

The regional roads have low presence and low pairing in the central positions, which confirms the appreciable asynchrony seen earlier. Local roads show some commonality with development but not otherwise in the central

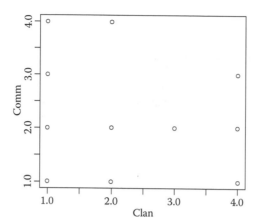

FIGURE 9.12
Occurrence by CLAN and (CLUMP) community clusters of 18 sentinels having RrdIVI in the highest cross-rank position. Some points are multiples.

positions, which is evidence that the last three signals are not redundant even though they have strong synchrony.

When there is apparent asynchrony, we can check for spatial specificity by isolating an aspect of it for plotting occurrence by CLAN and CLUMP community clusters. Accordingly, the 18 sentinels that have RrdIVI in the highest cross-rank position are considered to plot their occurrence by CLAN and CLUMP community clusters in Figure 9.12. Because these occurrences are distributed across CLAN and CLUMP community clusters, there is no pronounced spatial specificity.

Reference

Myers, W., and Patil, G. Preliminary prioritization based on partial order theory and R software for compositional complexes in landscape ecology, with applications to restoration, remediation, and enhancement. *Environmental and Ecological Statistics* **17**: 411–436, 2010.

10

Auto-Association: Local Likeness and Distance Decline

10.1 Introduction

Environments tend not to change in a fashion that is really random with movement across a local landscape. Such patterns of persistence and coherent change cause measurements made close together to have more similarity than those made farther apart, or so-called *spatial autocorrelation* (Bivand et al. 2008; Isaaks and Srivastava 1989; Webster and Oliver 2001). This is a major reason that maps help bring order to operations in landscape ecological analysis and regional planning. There is thus practical purpose in systematically capturing such coherence so that it is more easily exploited (Cressie 1993; Waller and Gotway 2004). However, these proximal propensities are also problematic for conventional statistical inference based on assumptions of information independence among observational instances (Schabenberger and Gotway 2008; Webster and Oliver 1990). In effect, a newly observed instance is only partially new if others occur in the same regional regime. Thus, there are effectively fewer degrees of freedom for assessing significance, which leads to stronger indications of significance than is warranted. Therefore, it may become necessary to appeal to pseudosignificance through Monte Carlo methods rather than running tests in the regular way. We explore this sort of spatial structure in the current chapter to support subsequent scenarios.

We begin by carrying forward nonparametric approaches in a multivariate mode. Thereafter, we turn toward more conventional considerations of spatial structure in single signals.

10.2 Cluster Companions

We return to the CLAN clustering that was done using four integrative indicators as sentinel signals in Chapter 5. Figure 10.1 is a map of these CLAN cluster numbers.

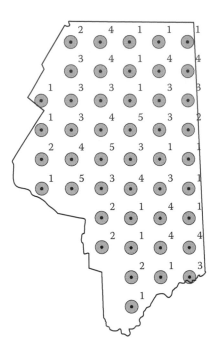

FIGURE 10.1

Map of CLAN cluster numbers based on four integrated vicinity indicators (IVIs) for Lackawanna County.

If there is local likeness at a scale of 2-km perimeters on a 5-km grid, then there should be a disproportionate number of neighboring sentinel sites that have the same CLAN cluster number. A relatively straightforward strategy is to tabulate the number of neighbors for each sentinel that belong to the same CLAN cluster, which we will call *cluster companions*.

Program 10.1 is a Python (2.7) program to count cluster companions within a specified distance or radial ring. Input is a space-delimited textual file with one line for each sentinel having an ID, *X* coordinate, *Y* coordinate, and CLAN number in that order. The program asks for file names, minimum and maximum distances, and a flag to use if there are no stations having the same cluster. The digit 0 should be used as a flag.

Program 10.1 Python (2.7) program to count cluster companions in radial ring.

```
# Counts like clusters in radial ring.
# Input is PtNum, X, Y, ClusNum
# Output is PtNum, X, Y, ClusNum, Companions
# Set No_such flag as 0
infil = raw_input("Input file?\n")
inpts = open(infil,"r")
npts = 0
```

```
inptz = []
nin = []
nopts = 0.0
cnts = []
cnt = 0
valstrs = []
for line in inpts:
    lining = line.split()
    ptnum = int(lining[0])
    ptx = float(lining[1])
    pty = float(lining[2])
    valu = int(lining[3])
    apt = (ptnum,ptx,pty,valu)
    inptz.append(apt)
    nin.append(nopts)
    cnts.append(cnt)
    valstrs.append(lining[3])
    npts += 1
feedbak = "Number of points" + " " + str(npts)
inpts.close()
print(feedbak)
done = 0
maxdis = raw_input("Maximum distance?\n")
maxdis = float(maxdis)
mindis = raw_input("Minimum distance?\n")
mindis = float(mindis)
nosuch = raw_input("No_such flag?\n")
if(maxdis <= 0.0):
    done = 1
while done < 1:
    indxx = 0
    for item in inptz:
        pnt = item[0]
        ptx = item[1]
        pty = item[2]
        vlu = item[3]
        cnts[indxx] = 0
        for itm in inptz:
            pntt = itm[0]
            ptxx = itm[1]
            ptyy = itm[2]
            vluu = itm[3]
            if pntt != pnt and vlu == vluu:
                dsqr = (ptx - ptxx) * (ptx - ptxx)
                dsqr += (pty - ptyy) * (pty - ptyy)
                dst = pow(dsqr,0.5)
                if dst >= mindis and dst <= maxdis:
                    nin[indxx] += 1
                    cnts[indxx] += 1
        indxx += 1
    indx = 0
    for item in nin:
        nopts = nin[indx]
        if nopts > 0:
            same = cnts[indx]
            addon = " " + str(same)
```

```
              valstrs[indx] += addon
        if nopts == 0:
              addon = " " + nosuch
              valstrs[indx] += addon
        indx +=1
    print("Set maximum distance = 0 to end\n")
    maxdis = raw_input("Maximum distance?\n")
    maxdis = float(maxdis)
    mindis = raw_input("Minimum distance?\n")
    mindis = float(mindis)
    if maxdis <= 0.0:
        done = 1
outfil = raw_input("Output file?\n")
outfile = open(outfil,"w")
indx = 0
for item in inptz:
    pnt = item[0]
    ptx = item[1]
    pty = item[2]
    vlu = item[3]
    outing = str(pnt) + " " + str(ptx)
    outing += " " + str(pty) + " "
    outing += valstrs[indx] + "\n"
    outfile.write(outing)
    indx += 1
outfile.close()
```

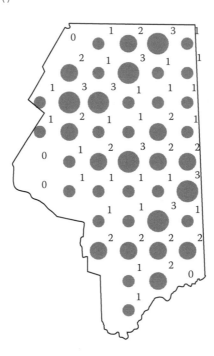

FIGURE 10.2
Map showing number of CLAN cluster companions among queen-move neighbors using proportional symbols with the number of companions as labels.

Figure 10.2 shows the cluster companion results for a radial distance of 8 km, which encompasses the queen-move neighbors of a sentinel station. The pattern of Figure 10.2 does not appear to be one of randomness. Because border stations have fewer neighbors, normalization by number of neighbors could be introduced as percent in current cluster. It is noted that all of the zeros in Figure 10.2 are border stations. Although we do not do so here, significance for nonrandomness could be tested by repeated random permutation of cluster ID numbers. **R** is well suited for conducting such permutation tests (Rizzo 2008).

10.3 Kindred Clusters

An avenue is available for extending the companion cluster idea to increase sensitivity beyond a binary match or no match. Clusters can be ordered in terms of similarity using centroids or representative ranks along the lines used earlier for precedence and progression, thus providing an extended "family" of kindred closeness for each cluster. Clusters for nearby stations can then be examined in terms of their similarity order, with lowest order number being most similar. A summation of the orders is obtained around each station. When these order sums are summed across all stations, they provide a statistic that is subject to permutation testing with a more diverse distribution than that for counts of cluster companions. With several strategies available for addressing cluster similarity, we do not make a specific choice at this juncture.

10.4 Local Averages

Rather than moving directly to full formalities of spatial statistics, we next compute local averages and compare the averages to actuals. If there is strong spatial structure of auto-association, there should be some similarity of actuals to averages over proximate positions. It might also be anticipated that similarity subsides with increasing spatial spans.

Program 10.2 is a Python (2.7) program that calculates averages for rings of distance around each station (locality) and is much like Program 10.1 in the way it is structured. This program takes a space-delimited textual file of data with the first item on each line being a station number, the second and third being X and Y coordinates, and the fourth being the value of a single sentinel signal (vicinity variate). The program calculates the average value for stations within a radial ring between a specified minimum distance and a maximum distance. The program asks for file names, minimum and maximum distances, and a flag to use if there are no stations in the radial ring.

Program 10.2 Python (2.7) program to calculate averages for stations in a radial ring.

```
# Computes average value in radial ring.
# Input is PtNum, X, Y, Value (space delimited)
# Output is PtNum, X, Y, Value, Average
Infil = raw_input("Input file?\n")
inpts = open(infil,"r")
npts = 0
inptz = []
nin = []
nopts = 0.0
sums = []
sum = 0.0
valstrs = []
for line in inpts:
    lining = line.split()
    ptnum = int(lining[0])
    ptx = float(lining[1])
    pty = float(lining[2])
    valu = float(lining[3])
    apt = (ptnum,ptx,pty,valu)
    inptz.append(apt)
    nin.append(nopts)
    sums.append(sum)
    valstrs.append(lining[3])
    npts += 1
feedbak = "Number of points" + " " + str(npts)
inpts.close()
print(feedbak)
done = 0
maxdis = raw_input("Maximum distance?\n")
maxdis = float(maxdis)
mindis = raw_input("Minimum distance?\n")
mindis = float(mindis)
nosuch = raw_input("No_such flag?\n")
if(maxdis <= 0.0):
    done = 1
while done < 1:
    indxx = 0
    for item in inptz:
        pnt = item[0]
        ptx = item[1]
        pty = item[2]
        vlu = item[3]
        for itm in inptz:
            pntt = itm[0]
            ptxx = itm[1]
            ptyy = itm[2]
            vluu = itm[3]
```

```
            if pntt != pnt:
                dsqr = (ptx - ptxx) * (ptx - ptxx)
                dsqr += (pty - ptyy) * (pty - ptyy)
                dst = pow(dsqr,0.5)
                if dst >= mindis and dst <= maxdis:
                    nin[indxx] += 1
                    sums[indxx] += vluu
        indxx += 1
    indx = 0
    for item in nin:
        nopts = nin[indx]
        if nopts > 0:
            avr = sums[indx]
            avr = avr / nopts
            addon = " " + str(avr)
            valstrs[indx] += addon
        if nopts == 0:
            addon = " " + nosuch
            valstrs[indx] += addon
        indx +=1
    print("Set maximum distance = 0 to end\n")
    maxdis = raw_input("Maximum distance?\n")
    maxdis = float(maxdis)
    mindis = raw_input("Minimum distance?\n")
    mindis = float(mindis)
    if maxdis <= 0.0:
        done = 1
outfil = raw_input("Output file?\n")
outfile = open(outfil,"w")
indx = 0
for item in inptz:
    pnt = item[0]
    ptx = item[1]
    pty = item[2]
    vlu = item[3]
    outing = str(pnt) + " " + str(ptx)
    outing += " " + str(pty) + " "
    outing += valstrs[indx] + "\n"
    outfile.write(outing)
    indx += 1
outfile.close()
```

Program 10.2 is now applied to the IVI that gives percentage for all kinds of development together. The first few lines of the input file are as follows.

```
1 450265.0626 4561637.415 0.99506
2 450265.0626 4566637.415 0.028636
3 455265.0626 4566637.415 1.911101
4 460265.0626 4566637.415 8.68448
5 445265.0626 4571637.415 0.035783
6 450265.0626 4571637.415 1.001932
```

The minimum distance is 0.0 m and the maximum distance is 5000.0 m. This includes only side neighbors (rook's move of chess). The "no_such flag" was specified as being –1.0 but does not come into play. The first few lines of output are as follows.

```
1  450265.0626  4561637.415  0.99506  0.028636
2  450265.0626  4566637.415  0.028636  1.30269766667
3  455265.0626  4566637.415  1.911101  5.467924
4  460265.0626  4566637.415  8.68448  4.966858
5  445265.0626  4571637.415  0.035783  0.515281
6  450265.0626  4571637.415  1.001932  2.18039225
```

We then switch to **R** for plotting the local average (last column) against the percentage development (next to last column) as shown in Figure 10.3. Perfect autocorrelation with these immediate neighbors would be reflected in such a plot as all points on a 45° line. This appears to be the situation for a group of stations having very low overall development, but shows departures for several others. It is instructive to fit a linear predictor to the data in Figure 10.3, and it provides for later comparison with a formal indicator of autocorrelation called Moran's I (Moran 1950). Figure 10.3 includes the prediction line having a slope of 0.2751 and 8.1938 as intercept. Three points are also labeled.

If the three labeled points in Figure 10.3 are eliminated, then the slope of the linear predictor changes to 0.7313, with 5.8070 as intercept. Because the slope more than doubles as a result of censoring 3 of 46 points, it is obvious that these are influential points. The usual statistical tests for linear predictor show this latter slope to be very highly significant, but it has already been stated that the usual tests are flawed in the presence of spatial autocorrelation.

The slope of the line for likeness of local average as computed above gives an overall indication of autocorrelation among the points. For effective mapping, however, we need an indicator that will reveal change in autocorrelation across the region of interest.

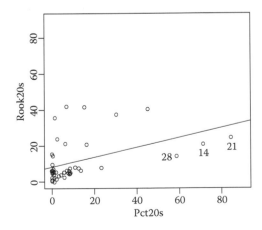

FIGURE 10.3

Plot of rook neighbors average (*Y*) against percent of all developments (*X*).

10.5 LISA: Local Indicator of Spatial Association

LISA is an acronym for "local indicator of spatial association," as put forward by Anselin (1995). There are three stipulations for a LISA (Schabenberger and Gotway 2005). First is that it speak to a particular locality. Second is that it be computable for each locality. Third is that aggregation of the local values be proportional to a global indicator of spatial autocorrelation.

We wish to pursue the relatively intuitive idea of local averages, but this does not conform to typical tactics. Therefore, we adapt an index of similarity due to Gower (1971) to suit the present purpose. The original index is intended to compare two instances based on the same variable, such that larger values of the index reflect greater likeness. The original form of the index is

$$1.0 - |X_I - X_J|/R_X$$

where X_I is the X value for the Ith instance, X_J is the X value for the Jth instance, and R_X is the range for the X variable. We adapt this as

$$1.0 - |X_I - Y_I|/R_X$$

where X is the value for a place, Y is the local average for the place, and R_X is the range for X over the data set. Inasmuch as Y is an average of X values, it must necessarily also lie within the range of X. Figure 10.4 shows this indicator plotted against sentinel station number (as "Index" on the horizontal axis) with a scatter of low values labeled. Figure 10.4 makes it evident that

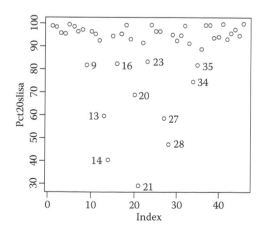

FIGURE 10.4
Indicator of spatial association using local averages plotted on vertical against station number on horizontal.

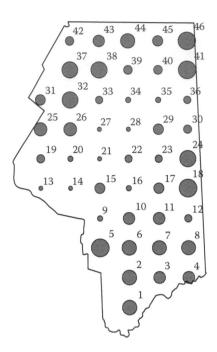

FIGURE 10.5
Decile map of local spatial association corresponding to Figure 10.4.

the labeled point positions are exceptions to a pattern of otherwise strong spatial association.

Figure 10.5 is a decile map corresponding to Figure 10.4. All of the labeled stations in Figure 10.4 except for number 9 are seen to lie in a diagonal strip of heavy development that is bordered by less developed land. Station 9 is also a queen-move neighbor to station 14 that is in the strip. Consequently, there is strong spatial pattern in weakness of association as well as in strength of association.

10.6 Picking Pairs at Lagged Locations

The study of auto-association can be extended by examining properties of pairs having specific separations in space. A specific separation under study is conventionally called a "lag," and it is often denoted by the letter "h." One might suppose that the "h" signifies a particular kind of "hop" to be taken from place to another place where potential pairings are sought. The specifics might limit the hop to one coordinate axis or the other, both axes, and/or even a particular direction. Alternatively, it may encompass a span of distances and/or directions. When a particular "h" has been examined and understood, the separations can be changed to explore progressive properties.

Program 10.3 provides for a pairing approach as a counterpart to the foregoing work with local averages. The program again takes a space-delimited textual file of data, with the first item on each line being a station number, the second and third being X and Y coordinates, and the fourth being the value of an integrative indicator. The program asks for file names and for minimum and maximum distances. The minimum and maximum distances again determine a ring around each station. A pairing is produced in the output file for each station that occurs in the ring.

**Program 10.3 Python (2.7) program for
pairing stations in a radial ring.**

```
# Program for pairings in a radial ring.
# Input is PtNum, X, Y, Value (space delimited)
# One output line for each pair as follows
# StationA ValueA StationB ValueB
Infil = raw_input("Input file?\n")
inpts = open(infil,"r")
npts = 0
inptz = []
for line in inpts:
    lining = line.split()
    ptnum = int(lining[0])
    ptx = float(lining[1])
    pty = float(lining[2])
    valu = float(lining[3])
    apt = (ptnum,ptx,pty,valu)
    inptz.append(apt)
    npts += 1
feedbak = "Number of points" + " " + str(npts)
inpts.close()
print(feedbak)
done = 0
maxdis = raw_input("Maximum distance?\n")
maxdis = float(maxdis)
mindis = raw_input("Minimum distance?\n")
mindis = float(mindis)
outfil = raw_input("Output file?\n")
outfile = open(outfil,"w")
for item in inptz:
    pnt = item[0]
    ptx = item[1]
    pty = item[2]
    vlu = item[3]
    for itm in inptz:
        pntt = itm[0]
        ptxx = itm[1]
        ptyy = itm[2]
        vluu = itm[3]
```

```
if pntt != pnt:
    dsqr = (ptx - ptxx) * (ptx - ptxx)
    dsqr += (pty - ptyy) * (pty - ptyy)
    dst = pow(dsqr,0.5)
    if dst >= mindis and dst <= maxdis:
        outing = str(pnt) + " " + str(vlu)
        outing += " " + str(pntt) + " "
        outing += str(vluu) + "\n"
        outfile.write(outing)
outfile.close()
ending = raw_input("Press ENTER to finish")
```

Program 10.3 is applied to the IVI that gives percentage for all kinds of development together. As before, the minimum distance is 0.0 m and the maximum distance is 5000.0 m, which encompasses only side neighbors (rook's move of chess). The first few lines of output are as follows, with the ID and value given for each pair; thus, the first line pairs station 1 and station 2.

```
1 0.99506 2 0.028636
2 0.028636 1 0.99506
2 0.028636 3 1.911101
2 0.028636 6 1.001932
3 1.911101 2 0.028636
3 1.911101 4 8.68448
```

The values for the pairs are plotted on an "h-scatterplot" (Isaaks and Srivastava 1989) in Figure 10.6. This type of plot is symmetric about the 45° line because each pair appears twice (permutation pairs) with the axes interchanged. Perhaps it may be helpful to think of this plot as a figurative bird with its tail at the origin and body oriented along the 45° line. Stations

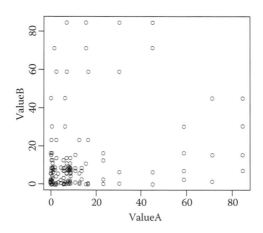

FIGURE 10.6
Five-kilometer h-scatterplot for percentage of all developments.

comprising the "wings" have already been identified in Figure 10.4 and mapped in Figure 10.5. Current interest for this figure centers on computing the *moment of inertia* about the 45° line (Isaaks and Srivastava 1989) as

$$\Sigma \ (V_A - V_B)^2/2n$$

where V_A is the value for the A member of the pair, V_B is the value for the B member of the pair, and n is the number of pairs. This moment of inertia is better known as a *variogram value* or more correctly as a *semivariogram value*. For data in Figure 10.6, 267.8875 is the value for the moment of inertia.

Note that the n in the denominator is the number of squares that are computed. Currently, this is 152 because permutation pairs are plotted. The same value would be obtained from combination pairs (76) if the denominator is adjusted accordingly, inasmuch as the permutation pairs just double the numerator as compared to combination pairs.

10.7 Empirical (Semi-)Variogram

The reason that the moment of inertia for the h-scatterplot above is called a variogram value is that major interest lies in how the variation among instances of the same variate changes with distance separating the instances. Intuitive anticipation would generally be that strength of autocorrelation might decline in some manner over a span of distances, and then perhaps stabilize when some distance of effective spatial independence is reached. Indeed, this will be the case for a stationary random process operating over space. As strength of autocorrelation declines, the variogram values increase because variation behaves opposite to association. Thus, a variogram plots variogram values against distance of separation (or *spatial lag*). For any given spatial lag, there is a corresponding h-scatterplot; Figure 10.6 shows the one for a 5-km lag distance. To obtain an (empirical) variogram, we need to plot the corresponding values against different lag distances.

It would be a protracted process to make an h-scatterplot for each lag and then put them together as a variogram. In practice, we do the reverse; that is, we make a variogram and then make h-scatterplots for lags that are particularly interesting. At this juncture, we use a contributed package for **R** from the CRAN library called *geoR* (by Paulo Ribeiro, Jr. and Peter Diggle) to make the variogram. The geoR package depends, in turn, on the "sp" package from CRAN. We will step through this in a bit of detail to clarify the considerations.

First is to load the library, make a data frame that fits the needs of geoR, and then convert it to the *geodata* class recognized in the package. As can be seen in what follows, the basic information for each geounit consists of an identifier, X and Y coordinates of a reference location, and the value of the variate under study.

```
> library(geoR)
Loading required package: sp
```

```
Analysis of geostatistical data
For an Introduction to geoR go to http://www.leg.ufpr.br/geoR
geoR version 1.6-32 (built on 2010-10—4) is now loaded
```

```
> head(Pct20sGeo)
   ID      PtX      PtY    Pct20s
1   1 450265.1 4561637 0.995060
2   2 450265.1 4566637 0.028636
3   3 455265.1 4566637 1.911101
4   4 460265.1 4566637 8.684480
5   5 445265.1 4571637 0.035783
6   6 450265.1 4571637 1.001932
> Pct20sGeo <- as.geodata(Pct20sGeo,coords.col = 2:3,data.col = 4)
> summary(Pct20sGeo)
Number of data points: 46

Coordinates summary
          PtX       PtY
min 435265.1 4561637
max 460265.1 4606637

Distance summary
      min       max
 5000.00 46097.72

Data summary
     Min.  1st Qu.    Median      Mean   3rd Qu.
0.71570  5.01400  10.72000  8.74300  84.53000
```

The next concern is to set up spans (or bins) of distance for the lags such that each will contain a reasonable number of pairs. This can be accomplished by a sequence of breaks. When accomplished in the following manner, there will be one less bin than the length of the sequence.

```
> Bins <- seq(0,46100,length=9)
```

The variogram can then be computed and plotted as shown in Figure 10.7.

```
> Varigrm <- variog(Pct20sGeo,breaks=Bins)
variog: computing omnidirectional variogram
> plot(Varigrm,type="b",main = "Pct20s Variogram")
```

It is helpful to assemble components of the variogram object in a tabular form.

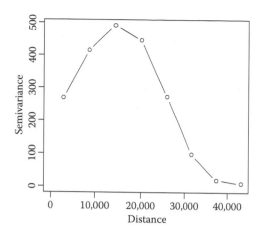

FIGURE 10.7
Omnidirectional empirical variogram for percentage of area in any kind of development.

```
> VarigrmTabl <- cbind(c(1:8),Varigrm$v,Varigrm$n)
> colnames(VarigrmTabl) <- c("Lag","Semi-variance","Pairs")
> VarigrmTabl
      Lag Semi-variance Pairs
[1,]   1    267.887557    76
[2,]   2    414.189026   237
[3,]   3    489.643972   175
[4,]   4    444.403734   237
[5,]   5    270.309116   141
[6,]   6     97.484308    95
[7,]   7     18.662595    60
[8,]   8      9.356803    14
```

Comparing the first lag to what was done with the h-scatterplot, it is to be noted that the number of pairs is given as combinatorial pairs rather than permutation pairs. Even with this level of binning, the last lag interval contains only 14 distinct pairs. For later reference, it is also important to note that this variogram is *omnidirectional*, meaning that the lag can occur in any direction.

The pattern of the plot for a stationary stochastic system would be one of initially rising to some level and then fluctuating about that level in a minor manner. The distance at which the level is reached is called the *range* and the level is called the *sill*. If there is an apparent nonzero intercept with the Y-axis, it is called a *nugget effect*. Although the empirical variogram in Figure 10.7 does have an initially rising pattern, it does not maintain its upper level but drops instead in a very pronounced manner. The decile map of the development variate shown in Figure 10.8 can help reveal what is going on in this regard. Figure 10.8 shows that the more centrally located stations tend to be similar in high development, and the peripheral stations tend to be similar in lower development. This is not surprising in view of the Wyoming Valley in

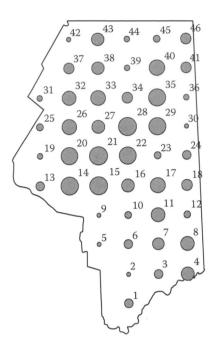

FIGURE 10.8
Decile map for percentage of area in all kinds of development, with labels for sentinel stations.

the central area surrounded by more mountainous terrain. Thus, the stations farthest apart also tend to be similar, which accounts for the deep decline after the variogram reaches a peak. Clearly, therefore, this does not conform to the pattern of a stationary stochastic system.

There is also a covariance function and a correlation function that show strength of auto-association versus lag distance. Both of these functions decline as the variogram increases. Both also involve a mean value for calculation. A particular virtue of the variogram is that it does not entail reference to a mean value.

10.8 Moran's *I* and Similar Spatial Statistics

We turn our attention now to the most commonly used indicator of spatial autocorrelation known as Moran's *I* (Moran 1950; Waller and Gotway 2004). In so doing, let us first revisit what was done above with averages of rook neighbors.

We created a composite of spatially selected neighbors, to which each selected neighbor contributed equally. An average of four can either be computed as the sum divided by 4, or by multiplying each one by 1/4 and summing the products. In the latter mode, each has a weight of 1/4 in forming the composite. The selection and compositing can be combined in weighting mode by giving rook neighbors the appropriate equal weights, with all other places having zero weight. We then computed a least squares slope coefficient for making predictions from the weighted composites. Moran's *I* is a natural extension of this mode to allowing all places a (weighted) contribution to the composite, with the weights usually being inversely proportional to the distance between places. Note that the weights for averaging totaled to unity (1.0), and this will effectively be true also for the general case through dividing the weighted sum by the sum of the weights.

With this as a preliminary, Moran's *I* can be formulated as

$$I = \{N / \Sigma_j (X_j - m_X)^2\} \, \{\Sigma_j \Sigma_k w_{jk}(X_j - m_X)(X_k - m_X)\}/W$$

where X_j and X_k are values at two places, m_X is the mean of X, w_{jk} is a weight for the pair of places, W is the sum of the weights over all pairs, and N is the number of places. The first term in braces divides the second term in braces by the variance of X (invert and multiply). The second term in braces is the weighted sum of products for deviations from the mean at pairs of places. This also has the general form of a weighted slope coefficient. Moran's *I* has an expected value of $-1/(N-1)$ in the absence of any spatial autocorrelation. Moran's $I = +1$ signifies positive (auto)correlation, and -1 signifies negative correlation.

GeoDa™ software (geodacenter.asu.edu), developed by Luc Anselin, works with shape files for spatial data exploration, including facilities for generation of spatial weights and determining Moran's *I* with testing by randomization. Suppression of selected points is also easy to accomplish. The application of GeoDa to the development data as above is shown in Figure 10.9. A standardization of data is automatically performed, but the result is

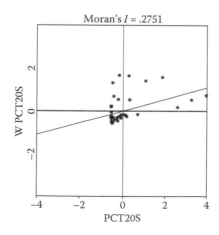

FIGURE 10.9
GeoDa representation of Moran's *I* for development data.

seen to be numerically the same (0.2751) as previously obtained more informally. When requested to do 999 randomizations for testing, a p-value of 0.0070 is reported, which is significant at the 0.01 level.

Geary's C (Geary 1954) is another statistic that can be used for practical purposes in assessing spatial autocorrelation. Using a parallel formulation to that for Moran's I given above, Geary's C can be cast as

$$C = \{(N-1)/\Sigma_j(X_j - m_X)^2\} \{\Sigma_j\Sigma_k w_{jk}(X_j - X_k)^2\}/2W$$

where the first term is essentially the same as above except that degrees of freedom are used instead of just N. Whereas the second term for Moran's I has the form of the numerator for a weighted slope, however, the second term for C has the form of a weighted variogram value, as discussed earlier in connection with Figure 10.6. Thus, the interpretation is different. Geary's C ranges between 0 and 2, with 1 signifying no autocorrelation. Values less than 1 signify positive autocorrelation, and values greater than 1 signify negative autocorrelation. Therefore, the interpretation of C is shifted and inverted relative to that for Moran's I.

References

Anselin, L. Local indicators of spatial association—LISA. *Geographic Analysis* **27**(2): 93–115, 1995.

Bivand, R., Pebesma, E., and Gomez-Rubio, V. *Applied Spatial Data Analysis with* **R**. New York: Springer, 2008.

Cressie, N. *Statistics for Spatial Data*. New York: Wiley, 1993.

Geary, R. The contiguity ratio and statistical mapping. *The Incorporated Statistician* **5**(3): 115–145, 1954.

Gower, J. C. A general coefficient of similarity and some of its properties. *Biometrics* **27**: 857–871, 1971.

Isaaks, E., and Srivastava, R. *An Introduction to Applied Geostatistics*. New York: Oxford Univ. Press, 1989.

Moran, P. Notes on continuous stochastic phenomena. *Biometrika* **37**: 17–33, 1950.

Rizzo, M. *Statistical Computing with* **R**. Boca Raton, FL: Chapman & Hall/CRC, 2008.

Schabenberger, O., and Gotway, C. *Statistical Methods for Spatial Data Analysis*. London: Chapman & Hall, 2005.

Waller, L., and Gotway, C. *Applied Spatial Statistics for Public Health Data*. Hoboken, NJ: Wiley, 2004.

Webster, R., and Oliver, M. *Statistical Methods in Soil and Land Resource Survey*. New York: Oxford Univ. Press, 1990.

Webster, R., and Oliver, M. *Geostatistics for Environmental Scientists*. Chichester, UK: Wiley, 2001.

11

Regression Relations for Spatial Stations

11.1 Introduction

A primary purpose for the localizing logic is to provide comparable cases for modeling methods. Regression approaches are among the more familiar modeling methods. This chapter is concerned with spatial structure in relation to regression. Inasmuch as we have seen strong suggestions of spatial trends throughout the preceding chapters, this can be our first focus with regard to regression.

11.2 Trend Surfaces

Trend surfaces are primarily useful and interpretable when they can be captured by relatively low-order polynomial functions of the coordinates. Trends of this nature provide important insights relative to prospects for using sophisticated spatial models based on underlying stochastic processes. A second-degree polynomial trend surface model can be configured readily in **R**.

Let us name a vector containing the X coordinates of sentinel stations as *PtX*, with the squares of the X coordinates being *PtXX*. Similarly, let Y coordinates be *PtY* and their squares be *PtYY*. Therefore, also, let products of X coordinates and Y coordinates be *PtXY*. Let us further use the vicinity variate for development (DvlpIVI) as the response under the name DvlpVV. Then a second-degree polynomial trend surface model is expressed as follows.

```
> DvlpTrnd <- lm(DvlpVV ~ PtX + PtY + PtXX + PtYY + PtXY)
```

A summary for this model is as follows.

```
> summary(DvlpTrnd)
```

```
Call:

lm(formula = DvlpVV ~ PtX + PtY + PtXX + PtYY + PtXY)
Residuals:
    Min      1Q  Median     3Q     Max
-20.372  -8.279  -3.977   3.345  60.980

Coefficients:
              Estimate Std. Error t value Pr(>|t|)
(Intercept) -9.094e+05  3.875e+05  -2.347   0.0240 *
PtX          1.576e-02  1.629e-01   0.097   0.9234
PtY          3.951e-01  1.631e-01   2.422   0.0201 *
PtXX        -9.417e-08  4.450e-08  -2.116   0.0406 *
PtYY        -4.381e-08  1.739e-08  -2.519   0.0159 *
PtXY         1.494e-08  3.332e-08   0.448   0.6563
---
Signif. codes:  0 '***' 0.001 '**' 0.01 '*' 0.05 '.' 0.1 ' ' 1

Residual standard error: 17.4 on 40 degrees of freedom
Multiple R-squared: 0.2222,     Adjusted R-squared: 0.125
F-statistic: 2.286 on 5 and 40 DF,  p-value: 0.06425
```

Regardless of the observed autocorrelation effects, the spatial structure is not well captured by a simple second-order polynomial trend surface. Even an overall *p*-value that is subject to inflation by spatial effects is not indicative of significance.

Nevertheless, it may be of interest to the curious to have maps showing percent of development, predicted percent development, and residuals. The map in Figure 11.1 shows percent development in 2% graduated steps, with labels giving percent development correct to one place after the decimal. The major feature appearing in Figure 11.1 is a diagonal ridge of development that is quite steep.

The predicted (fitted) values and residuals for the trend surface model are obtained as follows.

```
> DvlpTrndPrd <- fitted(DvlpTrnd)
> DvlpTrndRes <- residuals(DvlpTrnd)
```

The predicted values for the trend surface are mapped in Figure 11.2, and the residuals for the trend surface are mapped in Figure 11.3. The predicted values in Figure 11.2 do appear as a progressive trend, but the approximation to the original (Figure 11.1) is extremely poor. The trend version has negative values that are nonsensical, and the level rises to less than 30% of the actual maximum height.

The pattern of the residuals in Figure 11.3 actually looks more like the original pattern than the prediction pattern does. Thus, any thought of approximating the development pattern by a low-order polynomial trend surface can be set aside for this particular situation.

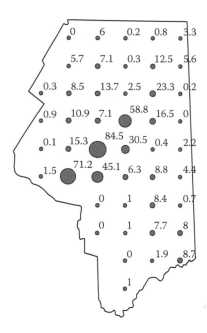

FIGURE 11.1

Percent development mapped with graduated symbols in 2% steps, with labels showing percent development to one place after the decimal.

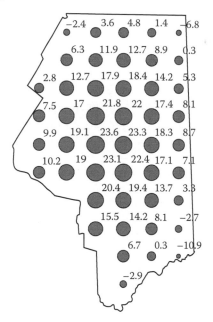

FIGURE 11.2

Percent development predicted trend surface as graduated symbols in 2% steps, with labels showing predicted value to one place after the decimal.

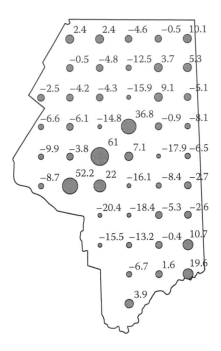

FIGURE 11.3
Residuals for trend surface as graduated symbols in 2% steps, with labels showing residual value to one place after the decimal.

11.3 Regression Relations among Vicinity Variates

Spatial regression is addressed in different disciplinary contexts (Bivand et al. 2008; Schabenberger and Gotway 2005). However, the analytical aspects are perhaps most accessible in social science and econometrics (Anselin 1988; Anselin 2002; Ward and Gleditsch 2008).

Ward and Gleditsch (2008) use **R** as their computing platform in a compact reference that is set in the context of social science. Anselin et al. (2006) provide an easily available software facility called GeoDa™ that is very approachable for exploratory work with good guidance.

The econometric and social science sectors emphasize two tactical modeling methods for dealing with dependence that is seemingly spatial. One modeling method sees distance dependent commonality and connectedness with interacting influences. This is called the *spatial lag* model because near neighbors are effectively considered to condition each other in an autoregressive mode. Accordingly, some weighted composite of values for neighbors will act as an independent variable in the regression. Thus, the influences of other independent variables are modulated by the spatial setting. A complicating conceptual consequence of this view (Ward and Gleditsch 2008) is to create a spatially based feedback system whereby perturbations propagate like rippling rings and then echo back and forth in a damping dance that seeks stability.

The other modeling method avoids the conceptual feedback by considering that the errors of estimation are spatially dependent rather than supposing that neighboring instances influence each other. This is called the *spatial error* model, and the task becomes one of modeling the spatially specific error to avoid inappropriate inference and inflated indications of significance.

Under both models, however, the ordinary least squares (OLS) method of estimation is not optimal and is replaced by other methods of maximum likelihood. OLS thus offers at best a "quick and dirty" first look that is not reliable for definitive declarations of significance.

We use GeoDa (geodacenter.asu.edu) to conduct regressions using the localized indicator for percent of all types of development (PCT20S or DvlpIVI) as the dependent variable, with road density indicators (RrdIVI and LrdIVI) as independent variables. A workbook for GeoDa (Anselin 2005) gives concise guidance along with a decision diagram for choosing among the different types of models. We use spatial weights computed on a limiting distance that encompasses only the rook neighbors for a station. Output for an OLS with spatial diagnostics is as follows.

```
REGRESSION
SUMMARY OF OUTPUT: ORDINARY LEAST SQUARES ESTIMATION
Data set           : Pts5kLacwn
Dependent Variable :       PCT20S   Number of Observations:   46
Mean dependent var :       10.7156  Number of Variables    :    3
S.D. dependent var :       18.399   Degrees of Freedom     :   43

R-squared          :     0.761672  F-statistic            :     68.7116
Adjusted R-squared :     0.750586  Prob(F-statistic)      :4.06708e-014
Sum squared residual:    3711.28   Log likelihood         :    -166.252
Sigma-square       :      86.3089  Akaike info criterion  :     338.505
S.E. of regression :      9.29026  Schwarz criterion      :     343.991
Sigma-square ML    :      80.68
S.E of regression ML:      8.98221
```

```
--------------------------------------------------------------------
   Variable    Coefficient    Std.Error    t-Statistic    Probability
--------------------------------------------------------------------
   CONSTANT      -12.9813      2.664935      -4.87115      0.0000154
     RRDIVI      3.689641      1.35475        2.723484      0.0092974
     LRDIVI      8.002511      0.8374076      9.556292      0.0000000
--------------------------------------------------------------------
```

```
REGRESSION DIAGNOSTICS
MULTICOLLINEARITY CONDITION NUMBER    3.910887
TEST ON NORMALITY OF ERRORS
TEST                   DF       VALUE         PROB
Jarque-Bera             2       55.56431      0.0000000
```

DIAGNOSTICS FOR HETEROSKEDASTICITY
RANDOM COEFFICIENTS

TEST	DF	VALUE	PROB
Breusch-Pagan test	2	18.7955	0.0000829
Koenker-Bassett test	2	5.534658	0.0628296

SPECIFICATION ROBUST TEST

TEST	DF	VALUE	PROB
White	5	9.907507	0.0778988

DIAGNOSTICS FOR SPATIAL DEPENDENCE
FOR WEIGHT MATRIX : **Pts5kLacwn.GWT**   (row-standardized weights)

TEST	MI/DF	VALUE	PROB
Moran's I (error)	-0.095985	-0.5664558	0.5710839
Lagrange Multiplier (lag)	1	0.8678900	0.3515401
Robust LM (lag)	1	3.2844412	0.0699394
Lagrange Multiplier (error)	1	0.6642896	0.4150496
Robust LM (error)	1	3.0808408	0.0792196
Lagrange Multiplier (SARMA)	2	3.9487308	0.1388494

COEFFICIENTS VARIANCE MATRIX

CONSTANT	RRDIVI	LRDIVI
7.101877	-2.042742	-0.862215
-2.042742	1.835348	-0.424245
-0.862215	-0.424245	0.701251

OBS	PCT20S	PREDICTED	RESIDUAL
1	0.99506	7.06238	-6.06732
2	0.02864	-6.99752	7.02616
3	1.91110	-0.25412	2.16523
4	8.68448	28.27625	-19.59177
5	0.03578	5.87484	-5.83906
6	1.00193	1.45027	-0.44834
7	7.69066	7.38458	0.30608
8	8.02262	10.42201	-2.39940
9	0.02863	-10.26436	10.29299
10	0.96649	6.54406	-5.57756
11	8.40072	14.55929	-6.15857
12	0.67988	2.89450	-2.21462
13	1.45980	-1.17236	2.63216
14	71.16982	33.93041	37.23941
15	45.07587	55.91841	-10.84255
16	6.33410	22.33752	-16.00342
17	8.76314	13.50351	-4.74037
18	4.38082	7.10468	-2.72386
19	0.05009	-0.02269	0.07277
20	15.33672	15.28953	0.04719
21	84.53446	69.49557	15.03889
22	30.45377	19.29506	11.15870
23	0.40787	2.14264	-1.73477
24	2.19725	2.92005	-0.72280
25	0.88150	-0.32625	1.20775
26	10.92744	23.63182	-12.70438
27	7.06312	12.61768	-5.55456
28	58.80332	49.19271	9.61061
29	16.47337	11.53116	4.94221
30	0.03592	2.89491	-2.85899
31	0.30043	-2.46784	2.76827
32	8.54793	23.09032	-14.54240

```
33      13.65686        11.34204         2.31483
34       2.51842         4.78574        -2.26731
35      23.31042        29.81130        -6.50087
36       0.18456        -8.21808         8.40263
37       5.73923         3.05824         2.68099
38       7.06514         7.03468         0.03046
39       0.26474         4.26265        -3.99791
40      12.54206         5.01404         7.52802
41       5.64759        11.17262        -5.52503
42       0.02147        -4.50041         4.52188
43       5.99814         3.90521         2.09293
44       0.17898         0.60850        -0.42952
45       0.82331         1.68857        -0.86526
46       3.32377        -4.90671         8.23048
========================= END OF REPORT =============================
```

All components of the regression (constant and coefficients for independent variables) are highly significant. The regression accounts for approximately 75% of the variation in the development indicator. The major spatial diagnostics (Moran's *I* for error, Lagrange Multiplier for lag, and Lagrange Multiplier for error) are all not significant, and the robust versions are not to be considered with this being so. Thus, there is no evidence of need to run either of the spatial models. However, we will run both of the spatial models anyhow for insights on lack of change in interpretation. Results for the spatial error model are as follows.

```
REGRESSION
SUMMARY OF OUTPUT: SPATIAL ERROR MODEL - MAXIMUM LIKELIHOOD ESTIMATION
Data set             : Pts5kLacwn
Spatial Weight       : Pts5kLacwn.GWT
Dependent Variable   :      PCT20S   Number of Observations:    46
Mean dependent var   :   10.715596   Number of Variables   :     3
S.D. dependent var   :   18.399037   Degree of Freedom     :    43
Lag coeff. (Lambda)  :   -0.169318

R-squared            :    0.767641   R-squared (BUSE)      : -
Sq. Correlation      : -              Log likelihood        : -165.874160
Sigma-square         :   78.659114   Akaike info criterion :    337.748
S.E of regression    :      8.869    Schwarz criterion     : 343.234245
```

```
----------------------------------------------------------------------
    Variable  Coefficient     Std.Error      z-value  Probability
----------------------------------------------------------------------
    CONSTANT    -13.38622      2.364914     -5.66034  0.0000000
      RRDIVI     3.620535      1.257625      2.878866  0.0039912
      LRDIVI     8.239333     0.7795715     10.56905  0.0000000
      LAMBDA    -0.1693183    0.1803614     -0.9387724  0.3478476
----------------------------------------------------------------------
```

```
REGRESSION DIAGNOSTICS
DIAGNOSTICS FOR HETEROSKEDASTICITY
RANDOM COEFFICIENTS
TEST                                   DF      VALUE        PROB
Breusch-Pagan test                      2    16.17448     0.0003074
```

```
DIAGNOSTICS FOR SPATIAL DEPENDENCE
SPATIAL ERROR DEPENDENCE FOR WEIGHT MATRIX : Pts5kLacwn.GWT
TEST                                 DF      VALUE        PROB
Likelihood Ratio Test                1       0.7566204    0.3843882

COEFFICIENTS VARIANCE MATRIX
   CONSTANT        RRDIVI        LRDIVI        LAMBDA
   5.592819      -1.709348     -0.699056     0.000000
  -1.709348       1.581621     -0.395622     0.000000
  -0.699056      -0.395622      0.607732     0.000000
   0.000000       0.000000      0.000000     0.032530


   OBS          PCT20S        PREDICTED        RESIDUAL        PRED ERROR
    1           0.99506         7.19733        -4.97404          -6.20227
    2           0.02864        -7.22536         7.03206           7.25400
    3           1.91110        -0.63123         1.84161           2.54233
    4           8.68448        29.07369       -20.39645         -20.38921
    5           0.03578         5.97096        -5.05455          -5.93518
    6           1.00193         1.27439        -0.42119          -0.27246
    7           7.69066         6.97095         0.44526           0.71970
    8           8.02262        10.65039        -3.85788          -2.62778
    9           0.02863       -10.64586         9.35339          10.67449
   10           0.96649         6.51850        -6.04887          -5.55201
   11           8.40072        14.52619        -6.61411          -6.12548
   12           0.67988         2.80541        -2.75716          -2.12553
   13           1.45980        -1.30692         5.89528           2.76672
   14          71.16982        34.47960        36.16683          36.69021
   15          45.07587        56.99602       -10.01684         -11.92016
   16           6.33410        22.34851       -16.47022         -16.01442
   17           8.76314        13.34890        -5.69778          -4.58576
   18           4.38082         6.81881        -2.85647          -2.43799
   19           0.05009        -0.21445         0.49380           0.26454
   20          15.33672        15.45654         1.46975          -0.11982
   21          84.53446        70.92055        13.34912          13.61391
   22          30.45377        19.16376        11.47960          11.29000
   23           0.40787         2.10053        -1.23118          -1.69266
   24           2.19725         2.90055        -1.09297          -0.70330
   25           0.88150        -0.53377         0.86786           1.41528
   26          10.92744        23.94385       -13.82167         -13.01642
   27           7.06312        12.56858        -5.01004          -5.50546
   28          58.80332        50.23115         8.93248           8.57217
   29          16.47337        11.57233         4.78648           4.90103
   30           0.03592         2.80943        -2.04407          -2.77350
   31           0.30043        -2.75223         1.91838           3.05266
   32           8.54793        23.36140       -15.00886         -14.81347
   33          13.65686        11.12271         1.59938           2.53415
   34           2.51842         4.69199        -2.14951          -2.17356
   35          23.31042        30.12263        -6.00700          -6.81221
   36           0.18456        -8.54199         7.88340           8.72655
   37           5.73923         2.92549         2.27369           2.81374
   38           7.06514         6.65570         0.57900           0.40944
   39           0.26474         3.99068        -3.49368          -3.72594
   40          12.54206         4.97373         6.86623           7.56833
   41           5.64759        11.00084        -3.95092          -5.35326
   42           0.02147        -4.81397         5.27545           4.83544
   43           5.99814         3.61442         2.66184           2.38372
   44           0.17898         0.49611        -0.43212          -0.31713
   45           0.82331         1.51844         0.19679          -0.69513
   46           3.32377        -5.22816         8.03988           8.55193
========================= END OF REPORT =============================
```

The constant and coefficients for independent variables are again highly significant, but the LAMBDA for spatial error is not. The regression again accounts for approximately 75% of variation in the dependent variable, with only a slight shift in the coefficients.

The spatial lag model is the other option. Results obtained by running that model are as follows.

```
REGRESSION
SUMMARY OF OUTPUT: SPATIAL LAG MODEL - MAXIMUM LIKELIHOOD ESTIMATION
Data set            : Pts5kLacwn
Spatial Weight      : Pts5kLacwn.GWT
Dependent Variable  :     PCT20S  Number of Observations:   46
Mean dependent var  :     10.7156 Number of Variables   :    4
S.D. dependent var  :     18.399  Degrees of Freedom    :   42
Lag coeff.  (Rho)   :     0.107983

R-squared           :     0.767119 Log likelihood        :    -165.804
Sq. Correlation     : -             Akaike info criterion :     339.607
Sigma-square        :     78.836   Schwarz criterion     :     346.922
S.E of regression   :     8.87897
```

```
-------------------------------------------------------------------
    Variable  Coefficient    Std.Error      z-value  Probability
-------------------------------------------------------------------
    W_PCT20S    0.1079827    0.1126217    0.9588096    0.3376547
    CONSTANT   -13.3201      2.57297     -5.176934     0.0000002
      RRDIVI    3.46526      1.326091     2.61314      0.0089715
      LRDIVI    7.776554     0.8226857    9.452643     0.0000000
-------------------------------------------------------------------
```

```
REGRESSION DIAGNOSTICS
DIAGNOSTICS FOR HETEROSKEDASTICITY
RANDOM COEFFICIENTS
TEST                               DF    VALUE        PROB
Breusch-Pagan test                  2    21.53134     0.0000211


DIAGNOSTICS FOR SPATIAL DEPENDENCE
SPATIAL LAG DEPENDENCE FOR WEIGHT MATRIX : Pts5kLacwn.GWT
TEST                               DF    VALUE        PROB
Likelihood Ratio Test               1    0.8976099    0.3434234
```

```
COEFFICIENTS VARIANCE MATRIX
   CONSTANT      RRDIVI      LRDIVI     W_PCT20S
   6.620175   -1.761316   -0.718049   -0.041103
  -1.761316    1.758516   -0.332946   -0.032265
  -0.718049   -0.332946    0.676812   -0.021450
  -0.041103   -0.032265   -0.021450    0.012684
```

```
    OBS        PCT20S      PREDICTED      RESIDUAL     PRED ERROR
      1        0.99506       5.33443      -5.12974      -4.33937
      2        0.028637     -7.29074       7.39325       7.31938
      3        1.9111       -0.22363       2.50811       2.13473
      4        8.6845       27.32676     -18.61183     -18.64228
      5        0.035783      4.56328      -4.98507      -4.52749
      6        1.0019        0.84950       0.19601       0.15243
```

7	7.6907	6.74120	1.11011	0.94946
8	8.0226	10.72126	-1.97370	-2.69864
9	0.028631	-8.29387	9.08837	8.32250
10	0.96649	6.32212	-4.97925	-5.35562
11	8.4007	13.94044	-5.23198	-5.53972
12	0.67988	3.12741	-2.07257	-2.44752
13	1.4598	-0.04054	-0.48754	1.50034
14	71.17	34.63215	36.96845	36.53767
15	45.076	56.47031	-12.55828	-11.39445
16	6.3341	23.21102	-16.54475	-16.87693
17	8.7631	13.30916	-3.82116	-4.54601
18	4.3808	6.55948	-1.91643	-2.17866
19	0.050089	-0.23059	0.25599	0.28068
20	15.337	17.41060	-3.13700	-2.07388
21	84.534	69.40582	15.47465	15.12864
22	30.454	21.46239	8.81682	8.99138
23	0.40787	2.66340	-2.47450	-2.25553
24	2.1973	2.48905	-0.05126	-0.29180
25	0.8815	-0.37792	1.61734	1.25942
26	10.927	23.49337	-11.94392	-12.56593
27	7.0631	15.40343	-8.75528	-8.34031
28	58.803	48.33349	10.44682	10.46983
29	16.473	12.51546	3.93274	3.95790
30	0.035925	2.28431	-2.64920	-2.24838
31	0.30043	-2.04113	3.02331	2.34156
32	8.5479	22.43443	-13.74787	-13.88651
33	13.657	11.31755	3.02042	2.33931
34	2.5184	6.27501	-3.87470	-3.75659
35	23.31	28.34427	-5.43917	-5.03385
36	0.18456	-7.28482	7.87273	7.46938
37	5.7392	2.98948	3.04644	2.74975
38	7.0651	6.31189	0.62360	0.75325
39	0.26474	3.66475	-3.51994	-3.40001
40	12.542	5.19828	7.67703	7.34377
41	5.6476	9.57139	-4.75724	-3.92380
42	0.021467	-4.86915	4.57404	4.89062
43	5.9981	2.88435	2.90697	3.11379
44	0.17898	0.07697	0.11185	0.10201
45	0.82331	0.81043	-0.55537	0.01288
46	3.3238	-5.01780	8.55272	8.34157

========================= END OF REPORT =================================

Note that the number of variables for the spatial lag model is listed as 4 instead of 3 because the lag term counts as a variable. That lag variable is not significant, but the constant and coefficients for road density variables are again highly significant. The coefficients are basically the same, as is also the percentage of variability accounted for by the regression. The likelihood ratio test for spatial dependence is not significant.

In summary, we find lack of spatial trend in percent development and find that road densities are effective indicators of development. There is no substantial evidence of residual spatial dependence after taking roads into account. One could conduct the usual analysis of larger residuals, but that would not differ from usual regression inquiry.

11.4 Restricted Regression

One can also entertain the idea that a regression relation may shift over a region. It would then be appropriate to extend the "local indicator of spatial association" concept by computing a separate (local) regression for each position while restricting the observations to some set of neighbors. The question then arises as to which neighbors should be used to obtain the spatially specific regression relation while retaining a sufficient number of observations. Because observations can be weighted in regressions, one way around this question is to weight the positions inversely according to distance so that more distant observations carry less weight. This strategy leads to geographically weighted regression (GWR) approaches (Fotheringham et al. 2002). Of course, an appropriate spatial regression model should be used in the presence of substantial spatial dependence. As for many other aspects of spatial analysis (Bivand et al. 2008), a contributed package (spgwr) is available in the CRAN library of **R** for conducting geographically weighted regression.

References

Anselin, L. *Spatial Econometrics: Methods and Models*. Dordrecht, The Netherlands: Kluwer Academic Publishers, 1988.

Anselin, L. Under the hood: issues in the specification and interpretation of spatial regression models. *Agricultural Economics* **27**: 247–267, 2002.

Anselin, L. *Exploring Spatial Data with GeoDa™: A Workbook*. Urbana, IL: Dept. of Geography, Univ. of Illinois, 2005.

Anselin, L., Syabri, I., and Kho, Y. GeoDa: An introduction to spatial data analysis. *Geographical Analysis* **38**: 5–22, 2006.

Bivand, R., Pebesma, E., and Gomez-Rubio, V. *Applied Spatial Data Analysis with* **R**. New York: Springer, 2008.

Fotheringham, A., Brunsdon, C., and Charlton, M. *Geographically Weighted Regression: The Analysis of Spatially Varying Relationships*. Chichester, UK: Wiley, 2002.

Schabenberger, O., and Gotway, C. *Statistical Methods for Spatial Data Analysis*. London: Chapman & Hall, 2005.

Ward, M., and Gleditsch, K. *Spatial Regression Models*. Los Angeles: Sage Publications, Inc., 2008.

12

Spatial Stations as Surface Samples

12.1 Introduction

In this chapter, we view the spatial signals from sentinels as surfaces that are sampled at the sentinel stations. In so doing, we expand our spatial purview to encompass two adjacent counties, which are Wyoming County and Luzerne County (Figure 12.1), both of which are involved in the Wyoming Valley context.

Strategic experience from previous chapters is utilized for structuring this more extensive surveillance setting. Accordingly, the previous sentinel stations for Lackawanna County are to be included in the locality layer. This requires more control of point positioning for a grid than is afforded by stock GIS facilities. Therefore, we provide a Python program (Program 12.1) for this purpose.

Program 12.1 Python program to generate grid of points.

```
# Generate grid of points.
ptx = raw_input("Lower-left X?\n")
ptx = float(ptx)
pty = raw_input("Lower-left Y?\n")
pty = float(pty)
rows = raw_input("Number of rows?\n")
rows = int(rows)
cols = raw_input("Number of columns?\n")
cols = int(cols)
stp = raw_input("Grid step?\n")
stp = float(stp)
npt = 1
outfil = raw_input("Output file?\n")
outfile = open(outfil,"w")
omits = " 1.#QNAN 1.#QNAN"
outfile.write("Point\n")
for ii in range(rows):
   ptxx = ptx
   for jj in range(cols):
      outing = str(npt) + " " + str(ptxx) + " " + str(pty) + omits + "\n"
      outfile.write(outing)
      npt += 1
      ptxx += stp
   pty += stp
outfile.write("END\n")
outfile.close()
ending = raw_input("Done -- press ENTER to close window.")
```

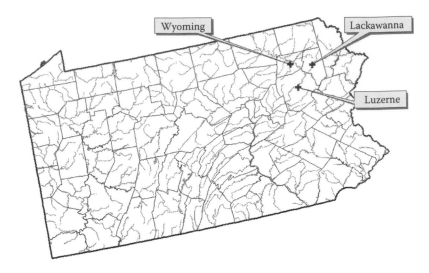

FIGURE 12.1
Tricounty area in Pennsylvania for the Wyoming Valley context. Major rivers are also shown.

256 257 258 259 260 261 262 263 264 265 266 267 268 269 270
241 242 243 244 245 246 247 248 249 250 251 252 253 254 255
226 227 228 229 230 231 232 233 234 235 236 237 238 239 240
211 212 213 214 215 216 217 218 219 220 221 222 223 224 225
196 197 198 199 200 201 202 203 204 205 206 207 208 209 210
181 182 183 184 185 186 187 188 189 190 191 192 193 194 195
166 167 168 169 170 171 172 173 174 175 176 177 178 179 180
151 152 153 154 155 156 157 158 159 160 161 162 163 164 165
136 137 138 139 140 141 142 143 144 145 146 147 148 149 150
121 122 123 124 125 126 127 128 129 130 131 132 133 134 135
106 107 108 109 110 111 112 113 114 115 116 117 118 119 120
91 92 93 94 95 96 97 98 99 100 101 102 103 104 105
76 77 78 79 80 81 82 83 84 85 86 87 88 89 90
61 62 63 64 65 66 67 68 69 70 71 72 73 74 75
46 47 48 49 50 51 52 53 54 55 56 57 58 59 60
31 32 33 34 35 36 37 38 39 40 41 42 43 44 45
16 17 18 19 20 21 22 23 24 25 26 27 28 29 30
1 2 3 4 5 6 7 8 9 10 11 12 13 14 15

FIGURE 12.2
Grid of 5-km point positions covering the tricounty area, with prior points in Lackawanna County having square symbols.

Points from this program are imported into the geographic information system (GIS) with the same geographic reference specifications as for the previous points in Lackawanna County. Figure 12.2 shows the resulting locality layer of point positions, with the previous Lackawanna County point positions being shown as squares. Points external to the tricounty area remain to be clipped.

With this close conformity of the grid to the area of interest, it would not be terribly tedious to identify the outlying points and selectively delete them. Renumbering of points can then be done as ranks for identifiers of those remaining. Clipping here has been done in the GIS, but renumbering has been done by exporting a copy of the attribute table and ranking remaining numbers in Excel©, with the result as shown in Figure 12.3.

Next is to use Program 8.1 for generating a localizing layer of octagonal proximity perimeters around the locality points. Figure 12.4 shows the octagonal vicinities (OCTIVs) around the locality points.

The final step in configuring the localizing layer is to clip the octagons to the boundary of the tricounty area, as shown in Figure 12.5.

The proximity perimeters in Figure 12.5 are ready for compilation of integrated vicinity indicators, and we will focus here on two primary components of land cover. One component is developed land as reflected in percentage of area having one of the 20s codes. The other component is forested land as reflected in percentage of area having one of the 40s codes. Practice thus far has been mostly to present maps of such indicators in terms of deciles.

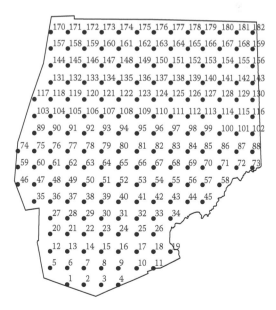

FIGURE 12.3
Grid of locality points in tricounty area with Lackawanna County sector.

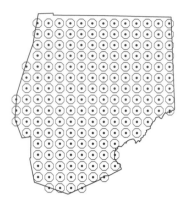

FIGURE 12.4
Octagonal proximity perimeters around locality points (sentinel stations) for tricounty area.

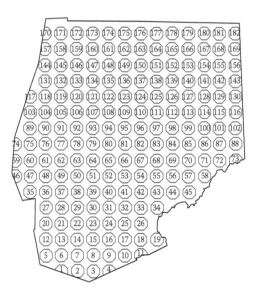

FIGURE 12.5
Clipped octagonal proximity perimeters for tricounty sentinel stations.

Accordingly, deciles for percentage of developed land (Pct20s) are shown in Figure 12.6 using graduated circular symbols.

Likewise, deciles for percentage of forested land (Pct40s) are shown in Figure 12.7 using graduated circular symbols. The diagonally oriented Wyoming Valley running from southwest to northeast is particularly evident as scarcity of forests (small symbols) in Figure 12.7. The same structure is shown in Figure 12.6, but it is not so clearly bounded.

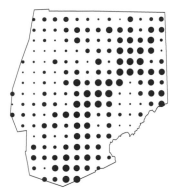

FIGURE 12.6
Deciles map of developed land IVI (Pct20s) using circular graduated symbols.

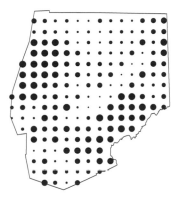

FIGURE 12.7
Deciles map of forested land IVI (Pct40s) using circular graduated symbols.

12.2 Interpolating Intensity Indicators as Smooth Surfaces

Any localized intensity indicator can be conceived as spot elevations on a surface, so intermediary places on that surface can be approximated by interpolation. A number of interpolation strategies are available for this purpose. In considering some such strategies, it should be kept in mind that the "actual" situation at any given locality could be obtained by selectively placing a sentinel station there. Thus, surface interpolation in the current context serves mainly for spatial synthesis and visualization.

When the purpose is synthesis and visualization, sophisticated surfacing strategies may not be substantially more suitable than simpler ones. A practical procedure is inverse distance weighting (IDW) of the nearby stations,

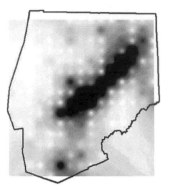

FIGURE 12.8
Inverse distance weighting (IDW) interpolation as 250-m cells of percent development (Pct20s), with light representing development and dark representing development.

which is a relatively simple weighted average (Isaaks and Srivastava 1989). Interpolation of the development indicator for 250-m cells using inverse squared distance is depicted in Figure 12.8, with lighter shades representing less development and darker shades representing more development.

Note how the dense development in the Wyoming Valley is now much more evident than it is in the deciles map of Figure 12.6. A surface depiction such as this is more effective when done in colors rather than grayscale.

A further aid to visualization comes by rendering the IDW development surface as a hill-shade perspective (2× vertical exaggeration) in Figure 12.9. Contours can likewise be derived from the IDW surface as shown in Figure 12.10 using a three-percent contour step. The sentinel stations can be shown in the contours without causing confusion, which would not be the case for the previous versions.

An IDW interpolation for percent forest (Pct40s) is shown in Figure 12.11 for comparison with Figure 12.8. This shows three major features, with absence

FIGURE 12.9
Hill-shade rendering of IDW development surface using a twofold vertical exaggeration.

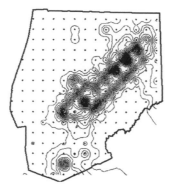

FIGURE 12.10
Contour rendering of IDW development surface using a three-percent step.

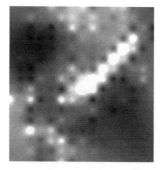

FIGURE 12.11
Inverse distance weighting (IDW) interpolation as 250-m cells of percent forest (Pct40s), with light representing less forest and dark representing more forest.

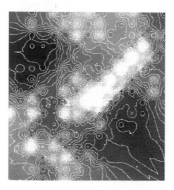

FIGURE 12.12
Inverse distance weighting (IDW) interpolation as 250-m cells of percent forest (Pct40s), with light representing less forest and dark representing more forest. White lines are three-percent contours.

of forest in the Wyoming Valley as the counterpart for presence of development. There are concentrations of forest in both the upper left (northwest) and lower right (southeast), which are mountainous areas where development is more difficult.

For surface depiction of the forest concentration, it can be helpful to add contours in white to help show structure in the heavily forested sectors as in Figure 12.12. The loss of contour information in the valley is actually advantageous because valley contours are more problematic for quick visual interpretation.

12.3 Spline Smoothing

In its mechanistic meaning, a spline is a connector. We began the previous chapter by attempting to approximate an intensity indicator by a polynomial

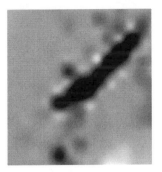

FIGURE 12.13
Spline interpolation as 250-m cells of percent development (Pct20s), with light representing less development and dark representing more development.

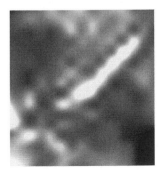

FIGURE 12.14
Spline interpolation as 250-m cells of percent forest (Pct40s), with light representing less forest and dark representing more forest.

trend surface model, which was not satisfactory in that particular situation. However, piecing together local polynomial parts gives much greater flexibility for analytical approximation. Spline strategies (Shabenberger and Gotway 2005) for smooth surfaces have several variants. A minimum curvature version in a GIS was used to produce Figure 12.13 for percent development and Figure 12.14 for percent forest. The spline scenarios tend to have a strong smoothing effect that lends emphasis to the major features of the surface and mutes the more minor features.

12.4 Kriging

The IDW and spline approaches need not necessarily match the actual values at the data points. In some situations, it would be desirable to have an exact interpolator with respect to the data points so that there are no residuals. More importantly, the interpolated values are estimates, and some situations require location-specific estimates of interpolation error. Such situations call for geostatistical approaches under the methodological umbrella of *kriging* (Webster and Oliver 2001). Kriging obtains optimal interpolation weights along with error estimates against a backdrop of stochastic random variables and stationarity. Kriging relies heavily on variogram models. Empirical variograms lend direction to variogram modeling, and an example of an empirical variogram was computed in Chapter 10. To many of those unfamiliar with it, variogram modeling can appear to be a combination of science and art that does not readily lend itself to "black-boxing." That is particularly true when there is evidence of directionality in the spatial variation, which is the case here due to the diagonally oriented Wyoming Valley. Nevertheless, some facilities do undertake a degree of automation, and we exercise one such facility here for comparative

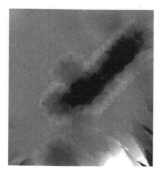

FIGURE 12.15
Interpolation by universal kriging with quadratic drift as 250-m cells of percent development (Pct20s), with light representing less development and dark representing more development.

FIGURE 12.16
Interpolation by universal kriging with quadratic drift as 250-m cells of percent forest (Pct40s), with light representing less forest and dark representing more forest.

purposes. Figures 12.15 and 12.16 are based on choice of universal kriging with quadratic drift. The perceived patterns in these figures are similar to the previous patterns, but there is a strong imprint of the parent point pattern along with notable edge artifacts.

Kriging, in particular, is an extensive and specialized subject with a substantial history, which we will not pursue further here. The three previous references along with Bivand et al. (2008) provide general treatments and can serve as an entryway to the larger literature and software systems.

References

Bivand, R., Pebesma, E., and Gomez-Rubio, V. *Applied Spatial Data Analysis with* **R**. New York: Springer, 2008.

Isaaks, E., and Srivastava, R. *An Introduction to Applied Geostatistics*. New York: Oxford Univ. Press, 1989.

Schabenberger, O., and Gotway, C. *Statistical Methods for Spatial Data Analysis*. London: Chapman & Hall, 2005.

Webster, R., and Oliver, M. *Geostatistics for Environmental Scientists*. Chichester, UK: Wiley, 2001.

13

Shifting Spatial Structure

13.1 Introduction

This penultimate chapter serves to focus interest for surveillance on changes in spatial structure. We examine possibilities for several strategies that are adaptations of approaches introduced in earlier chapters. Scan statistics, as considered in Chapter 7, provide the point of beginning.

13.2 Space–Time Hotspots

The volume edited by Glaz et al. (2009) encompasses consideration of various aspects and applications of scan statistics, including detection of significant space–time shifts in occurrence. The extension from map space to space–time is relatively straightforward when time is treated as a third dimension. The SaTScan facility by Kulldorff (http://www.satscan.org), as applied in Chapter 7, makes the extension by scanning cylindrical regions of a three-dimensional space instead of circles in a two-dimensional space. Having illustrated applications for different statistical models in space, we refer to the examples distributed with SaTScan for incorporating time.

13.3 Salient Shifts

When surveillance is conducted for the same set of signals on two occasions or with two different supports, a scatterplot of salient sequences will reveal shifts in spatial structure. We explore this strategy here in terms of different supports by extracting the Lackawanna County subset from the tricounty area encompassing the Wyoming Valley. Localization for this subset uses octagons having 2-km circumscribing radii, whereas previous work for

Lackawanna County alone used mostly circle-based proximity perimeters. The Lackawanna County subset can be extracted easily by clipping. Figure 13.1 shows the clipped subset with ID numbers from the larger area.

We will use percent development (Pct20s) and percent forest (Pct40s) as a bivariate set of signals to be compared for octagonal support and circular support. In so doing, we recall that the octagonal support encompasses 90% of the circular support. However, precedence plots and salient sequencing require a set of compatible indicators. Development and forest are opposing indicators because one decreases as the other increases and because development indicates human influence whereas forest indicates naturalistic condition. Therefore, we need to couple percent development and percent nonforest to obtain a pair of compatible indicators with mutual sense of indication. Accordingly, we proceed to compute precedence and salient sequencing for both circles and octagons. We then make a scatterplot of these two salient sequences as shown in Figure 13.2.

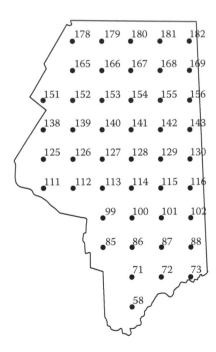

FIGURE 13.1
Lackawanna County subset of stations from larger tricounty area encompassing the Wyoming Valley.

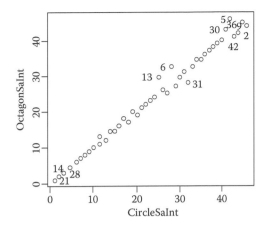

FIGURE 13.2
Plot of salient sequence for development and nonforest in octagons versus salient sequence for development and nonforest in circles.

The locations for heaviest development and nonforest in both circles and octagons are as follows, with 21 being most so and 14 next.

```
   POINT_X POINT_Y    Pct20s Pct20sCir  ID PctNon40s PctNon40sCir
14  440265 4581637 71.72271  71.16982  14  84.72852     84.42870
21  445265 4586637 86.21183  84.53446  21  90.64090     89.46540
28  450265 4591637 60.00159  58.80332  28  71.58070     70.16891
```

Stations 6 and 13 are intermediate for salient sequence but show some small inconsistency between circles and octagons. If these were two different time periods, then some slight shifting of spatial structure relative to other stations would be indicated.

Since positions of shifts are important, a map should be prepared. Relative size of shifts should be shown in the map, but these can be misleading if absolute sizes of shifts are not also provided. Our solution to this is to use actual shifts as labels with proportional symbols as shown in Figure 13.3. The shift numbers shown are obtained by subtracting salient sequence numbers for circles from salient sequence numbers for octagons. The sizes of the symbols are proportional to the absolute value of the salient sequence shift.

It may also be desirable to make micromaps (Carr 2010) in the proximity perimeters of the stations that exhibit substantial amounts of change. The micromaps can be hot-linked to their locations in the more general map.

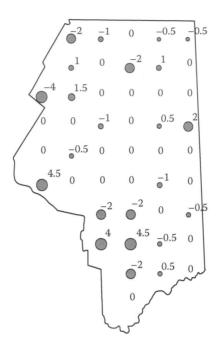

FIGURE 13.3
Size of salient shifts (octagons – circles) as symbols proportional to the absolute value of difference.

13.4 Paired Plots

If the number of indicators is relatively few, then a lattice of paired plots is helpful in detecting discrepancies that occur between sets, as illustrated in Figure 13.4.

Figure 13.4 shows that the sets are well matched, with percent development and percent nonforest in circles being like those in octagons. However, plots of development versus nonforest reflect the influence of other open lands such as grassland and cropland.

13.5 Primary Partition Plots

We have emphasized components of land cover as being important indicators of status for the human environment interface and among those for which spatial data are most readily available. The balance between development and forest is particularly informative for temperate and tropical regions aside from natural grasslands and savannahs in semiarid regions.

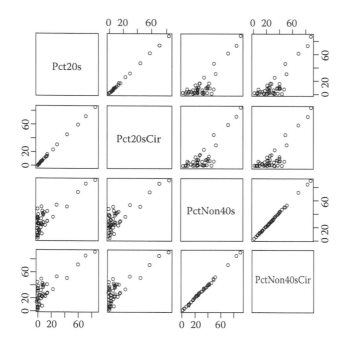

FIGURE 13.4

Lattice of paired plots for percent development and nonforest in octagons and circles. The corresponding indicators between sets are very well matched, but influence of other nonforest besides development is evident within sets.

We put forward an adaptation of our precedence plots for visualization of such balance in a comparative manner among locations. We plot the percentage of forest on the vertical side and the percentage of development on the horizontal side, as shown in Figure 13.5. Heavily forested areas with little development will appear in the top of the plot, and heavily developed areas with little forest will appear in the toe of the plot. Extent of offset to the left of the diagonal line indicates substantial component(s) of cover as nonforest other than development. We also bring forward a map of station ID numbers in Figure 13.6 so that the labeled stations can be located. The five labeled stations are seen to constitute a spatial cluster in the Wyoming Valley. These sorts of plots can be used over time to track spatial shifts in degree of human influence and occurrences of development encroaching on and fragmenting forests.

The utility of plots such as this can be extended to any context involving percentages where there is a primary pair of counterpart conditions. Let one component be considered the *positive part* and the other as the *contrary component*. Proceeding to plot the positive part on the vertical and the contrary component on the horizontal will provide a visual for comparative analysis of compositional condition.

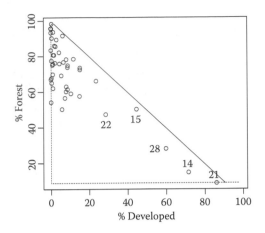

FIGURE 13.5
Plot of cover condition for stations as percent forest versus percent developed.

FIGURE 13.6
Station numbers in Lackawanna County.

13.6 Backdrop: Spectral Detection of Change with Remote Sensing

Most change detection work involves remote sensing either directly or indirectly (Aspinall and Hill 2008; Canty 2010; Chuvieco and Huete 2010; Wang 2010; Weng 2011). In the backdrop to Chapter 6 (Section 6.5), it was briefly outlined that most land cover mapping is accomplished primarily by determining spectral signatures for land cover classes. When land cover maps become available in this manner, they can then be analyzed by the localization methods that we have set forth to determine compositional differences from one time to another.

Detecting occurrence of change without making a second land cover map can be approached in a different manner. When multiband images having the same or similar spectral bands are available for different time periods, a relatively straightforward "change vector" approach can be used. This begins by pairing each band from one period with its counterpart band in another period and then computing a difference band for each pair. The difference bands are then brought together to form a multiband difference image. Each of the difference bands becomes a component of a multidimensional spectral change vector. Various composite renditions of the change vectors are then created to lend emphasis (enhancement) to different types of changes. Each type of change having a characteristic composite appearance is then explored to develop an etiology of change for the image, and concentrations of occurrence are mapped.

A hybrid spectral/spatial approach (Myers and Patil 2006) is to overlay spectral patterns from a later image on a raster land cover map from an earlier date in order to develop a cross-walk of spectral signatures from the later image with classes from the earlier map. Occurrences of spatial anomalies from the matched patterns are then flagged as areas of interest for investigation of change. Other kinds of similarly sophisticated spectral/spatial approaches are continually being explored in the realms of remote sensing research.

Micromaps can also be made within proximity perimeters by on-screen interactive delineation using a mouse as a cursor, with an image overlay such as that shown earlier in Figure 6.17.

References

Aspinall, R., and Hill, M., Eds. *Land Use Change: Science, Policy and Management*. Boca Raton, FL: Taylor & Francis/CRC, 2008.

Canty, M. *Image Analysis, Classification, and Change Detection in Remote Sensing*, 2nd ed. Boca Raton, FL: Taylor & Francis/CRC, 2010.

Carr, D. *Visualizing Data Patterns with Micromaps*. Boca Raton, FL: Taylor & Francis, 2011.

Chuvieco, E., and Huete, A. *Fundamentals of Satellite Remote Sensing*. Boca Raton, FL: Taylor & Francis/CRC, 2010.

Glaz, J., Pozdnyakov, V., and Wallenstein, S., Eds. *Scan Statistics: Methods and Applications*. Boston: Birkhauser, 2009.

Myers, W., and Patil, G. P. *Pattern-Based Compression of Multi-Band Image Data for Landscape Analysis*. New York: Springer, 2006.

Wang, Y., Ed. *Remote Sensing of Coastal Environments*. Boca Raton, FL: Taylor & Francis/CRC, 2010.

Weng, Q., Ed. *Advances in Environmental Remote Sensing: Sensors, Algorithms and Applications*. Boca Raton, FL: Taylor & Francis/CRC, 2011.

14

Synthesis and Synopsis with Allegheny Application

14.1 Introduction

This final chapter is intended to be essentially self-standing so that those having preliminary interest can read the first and last chapters to obtain a basic appreciation of the intention and approach that have been the main thrust throughout the book. However, the other chapters should be read in numerical order thereafter because each builds upon those preceding. Much of this chapter appeared in the electronic proceedings for the 2011 Joint Statistical Meetings (JSM) under submission number 300319 for a presentation by the same title as this book (Myers and Patil 2011).

14.2 Localization Logic

We set forth a strategy for spatial surveillance based on localizing layers of spatial data from geographic information systems (GIS) and supplemental sources to obtain integrated vicinity indicators (IVI) amenable to multivariate and multiscale analysis. We use Allegheny County in Pennsylvania, U.S.A., encompassing the city of Pittsburgh as a context for explication. Figure 14.1 shows generalized land cover from the National Land Cover Database (NLCD) produced by the Multi-Resolution Land Characteristics (MRLC) consortium (Chander et al. 2009; Homer et al. 2001) as a backdrop for the county. Figure 14.2 shows an enlargement of the rectangle marked in Figure 14.1. One of the notable physiographic features of Pittsburgh is the confluence of the Allegheny River and the Monongahela River to form the Ohio River.

The initial step in our surveillance strategy is to generate a square pattern of points spanning the extent of interest. This pattern of points comprises

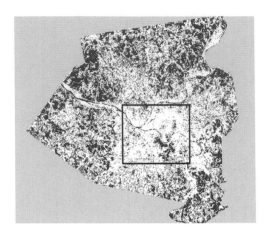

FIGURE 14.1
Allegheny County, PA, showing generalized land cover in shades of gray. Black is water, and white is developed area. Shades of gray show from darker to lighter as forest, shrubland and transitional, grasses, and cropland.

a *locality layer*. The next step is to generate a polygonal proximity for each point, with the collection of such polygons constituting a *localizing layer*.

An IVI is then synthesized (Myers and Patil 2006) within the proximity perimeters for each characteristic to be considered and is attributed to the points as properties. A point pattern with one or more such properties constitutes a *poly-place purview* as a subject for statistical surveillance. When stochastics of models are imposed, this becomes a *poly-place process purview*. Surveillance can include cluster configurations, prioritizations

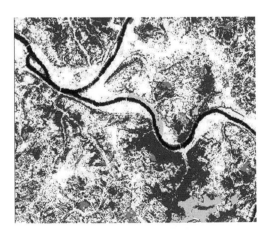

FIGURE 14.2
Enlargement of rectangle marked in Figure 14.1.

by partial order theory, hotspot detection with scan statistics, spatial regression relations, etc.

14.3 Locality Layer

The first order of operations is to construct the locality layer. This begins with determining the bounding box for the area of interest (Allegheny County), which has X and Y coordinate limits (in meters) as follows.

$$2079570$$
$$1305127 \qquad\qquad 1369814$$
$$2017115$$

Next is to select spacing for the points, which in this instance is to be 5 km (5000 m). A rectangular array of points can then be generated that spans the bounding box, as shown in Figure 14.3.

Points lying within the county are then extracted and renumbered in sequence by assigning rank numbers for IDs to the remaining points, as shown in Figure 14.4.

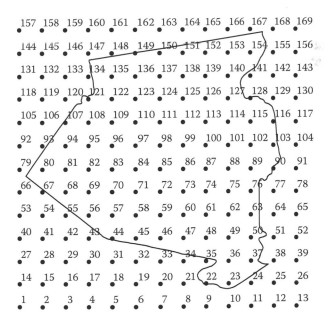

FIGURE 14.3
Rectangular array of points spanning the bounding box for the county.

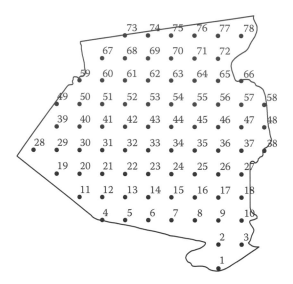

FIGURE 14.4
Seventy-eight sequentially numbered points in Allegheny County.

14.4 Localizing Layer

The second operation is to put polygonal proximities around the locality points as a localizing layer. IVIs are then compiled within these proximity perimeters.

The tendency of GIS analysts would be to put circular buffers around the points (Burrough and McDonnell 1998; Chrisman 2002; Demers 2009). However, the many chords used to approximate circles do not make efficient use of computer storage space for purposes of a localizing layer. An octagon inscribed in a circle encompasses approximately 90% of the area of the circle while using only eight chords. Therefore, octagons are advantageous for this purpose. A Python (Lutz 2009; Summerfield 2009) program has been written to take a textual file of point identifiers and coordinates for the locality layer as input and generate an octagon around each point for importing into GIS. Octagons for 2-km-radius circumscribing circles are shown in Figure 14.5.

It is also necessary to clip octagons straddling the border in order to avoid edge bias in compiling IVIs, as shown in Figure 14.6. A localizing layer of 1-km-radius octagons is likewise constructed for later exploration of sensitivity to scale of support.

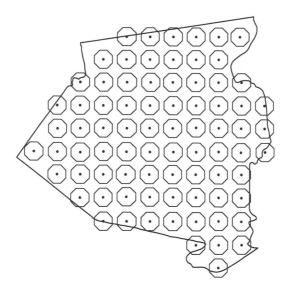

FIGURE 14.5
Octagons with 2-km radius around locality points for Allegheny County.

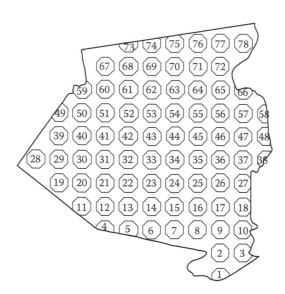

FIGURE 14.6
Clipped octagons with 2-km circumscribing radii around locality points for Allegheny County.

14.5 Poly-Place Purviews

At this juncture, we use GIS facilities of ArcGIS© (Ormsby et al. 2010) to tabulate areas of different land cover types within the respective octagonal proximities and compile percentages of selected land cover types as IVIs. The present poly-place purview for a 2-km radius is composed of percent development, percent forest, and percent water. A deciles map for percent development is shown in Figure 14.7.

The poly-place purview is transferred to **R** software (Allerhand 2011; Hogan 2010; Horton and Kleinman 2011; R Development Core Team 2008; Venables et al. 2005) in tabular form, with the head (leading lines) given in Table 14.1. Sid is the sequential ID, and the other names should be

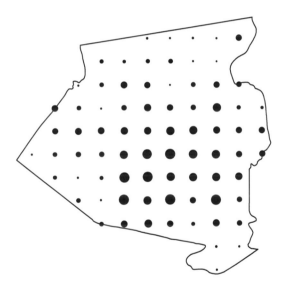

FIGURE 14.7
Deciles map of percent development for 2-km poly-place purview of Allegheny County.

TABLE 14.1

Leading Lines (Head) of 2-km Poly-Place Purview of Allegheny County

Sid	POINT_X	POINT_Y	PctDevl	PctFor	PctWatr
1	1352624	2024615	3.704869	66.40428	2.581863
2	1352624	2029615	4.375497	61.86158	0.000000
3	1357624	2029615	1.422963	82.20477	0.073601
4	1327624	2034615	9.981933	52.42397	0.000000
5	1332624	2034615	27.472629	51.53090	0.000000
6	1337624	2034615	42.584341	44.20751	0.000000

self-explanatory. A corresponding 1-km purview is also developed. Since these two purviews share the same set of localities, they can also be easily combined.

14.6 Significant Spatial Sectors with Scan Statistics

Scan statistics (Wallenstein 2009) are well suited to finding contiguous clusters with increased incidence of influences according to relative risk for our situation. For present purposes, we use SaTScan™ (www.satscan .org) software (Costa and Kulldorff 2009; Kulldorff 1997). In its primary modes, SaTScan considers point-referenced area data. Expanding circles or ellipses are used to search for the most likely contiguous cluster, and Monte Carlo methods are then used to assess significance.

We proceed to use SaTScan software with developed land being like an infective influence to explore elevated extents in this regard. We choose a Bernoulli model, with 30-m cells of the land cover data being developed or not. Very highly significant hotspot locations occupy the center of the county, with one solitary hotspot on the periphery, as shown in Figure 14.8.

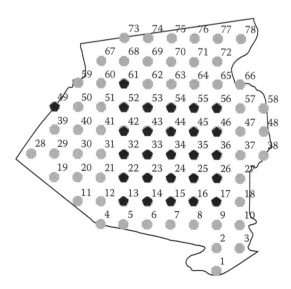

FIGURE 14.8
Hotspot locations of developed land in Allegheny County from SaTScan.

14.7 Scale Sensitivity and Partial Precedence

We have two poly-place purviews for the same set of localities but different octagonal support areas. The foregoing has pertained to octagonal support having a 2-km circumscribing radius. Octagonal support having a 1-km radius is also available. For two localizing layers, the simplest method of showing sensitivity to scale of support is to graph the corresponding intergrative vicinity indicators as a scatterplot, as shown in Figure 14.9.

The differences in percent development between 2-km support and 1-km support for the localities identified in Figure 14.9 are given in Table 14.2.

More generally, partial (percent) pairwise precedence (Myers and Patil 2010) provides comparative capability for several supports and also serves prioritization purposes involving several IVIs. Consensus and conflicts among indicators are considered concurrently and shown schematically. When the indicators differ only in scale of support, consensus reflects lack of sensitivity to scale, and conflict reflects sensitivity to scale. Giving the indicators same sense of sign is a preliminary to the process, which already pertains in scale of support scenarios.

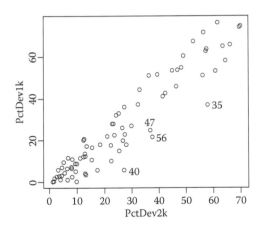

FIGURE 14.9
Scatterplot of percent development with 1-km radius support (Y) versus percent development with 2-km radius support (X) and selected points identified.

TABLE 14.2

Differences due to Support for the Points That Are Identified in Figure 14.9

Sid	POINT_X	POINT_Y	PctDev2k	PctDev1k
35	1347624	2049615	57.97470	36.870618
40	1322624	2054615	27.18308	5.536111
47	1357624	2054615	36.89737	24.721603
56	1352624	2059615	37.62928	21.578612

Suppose that a protocol is provided for prescribing precedence or lack thereof between members of a pair of informational instances. The protocol determines that either (1) one member of the pair has precedence over the other or (2) this pair of instances is indefinite with respect to precedence according the indicators. For a particular pair, we will thus have the following:

(1) Instance *i* has *pairwise precedence* (pp) over instance *j* with the reciprocal relation being one of *definite deficiency* (dd).

(2) Instances *i* and *j* are pairwise *indefinite instances* (ii) for which precedence is not assigned by the protocol.

Each of the instances can then be treated as focus of comparison by pairing with each of the other cc = *n* − 1 *competing cases* (instances) to tabulate the following, where *ff*() denotes the focal frequency of pairwise occurrence.

$$PP = 100 \times ff(pp)/cc$$

$$DD = 100 \times ff(dd)/cc$$

$$II = 100 \times ff(ii)/cc$$

PP is thus the *precedence percent* and DD + II is the *deficiency domain*, with DD being the definite part of that domain. For the present purpose, precedence is prescribed by the so-called *product–order protocol* (Bruggemann and Patil 2011). In order for instance *i* to have precedence over instance *j*, *i* must be better in at least one respect and cannot be worse in any respect.

We then produce a *precedence plot* with PP on the vertical against DD on the horizontal, as in Figure 14.10. Points plotting close to the diagonal line

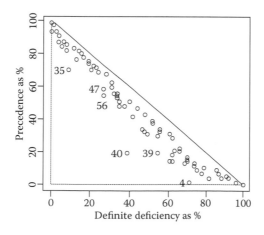

FIGURE 14.10
Precedence plot for percent development at two scales of support, with selected localities identified.

in Figure 14.10 have little sensitivity to scale of support, whereas points farther to the left are more scale sensitive. Points in the upper-left corner have consensus on high development, whereas point in the lower-left have consensus on low development. Even with only two indicators, Figure 14.10 is more informative than Figure 14.9 because point number 4 is more evidently sensitive. It is also easier to order the file of points by PP or DD than to order jointly according to Figure 14.9.

14.8 Cluster Components and Cluster Companions

Study of spatial structure in a multivariate mode can commence with clustering (Abonyi and Balaz 2007; Xu and Wunsch 2009). The dendrogram for hierarchical clustering of the three 2-km land cover indicators is shown in Figure 14.11. From this dendrogram, five clusters appear interesting. Locations of the five clusters are mapped in Figure 14.12, and the cluster frequencies are given in Table 14.3.

Cluster centroids are plotted in Figure 14.13, with numeric values given in Table 14.4.

Except for clusters 1 and 2, the clusters are well separated with regard to percent development. The first two clusters both have low percentage development but differ with regard to forest cover. The alternative to development, forest, and water is "open" land having agriculture, grasses, etc. Therefore, cluster 1 has more of such open land.

For visualization of multivariate spatial autocorrelation (Lloyd 2011), we can use cluster companions as percent of neighbors that belong to the same cluster. This is depicted in Figure 14.14 using graduated symbols and queen-move neighbors.

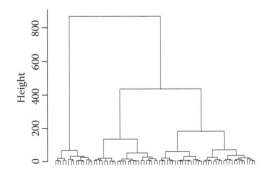

FIGURE 14.11
Dendrogram for hierarchical clustering of 2-km land covers by Ward linkage.

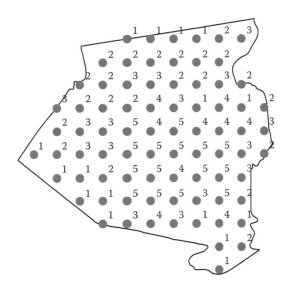

FIGURE 14.12
Localities of five clusters based on 2-km land cover in Allegheny County.

TABLE 14.3

Frequencies for Five Clusters Based on 2-km Land Covers

Cluster:	1	2	3	4	5
Frequency:	17	21	16	9	15

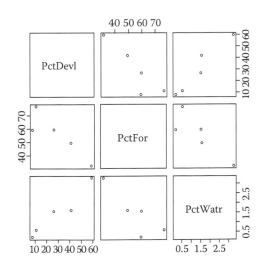

FIGURE 14.13
Lattice of paired plots of land cover percentages for cluster centroids.

TABLE 14.4

Centroid Values for Five Clusters Based on
2-km Land Covers

	PctDevl	PctFor	PctWatr
1	7.694325	59.11081	0.1911984
2	10.638389	76.63932	0.5497448
3	26.516920	59.41316	1.5312111
4	41.478171	49.20947	1.5752974
5	59.542506	32.01061	3.2798007

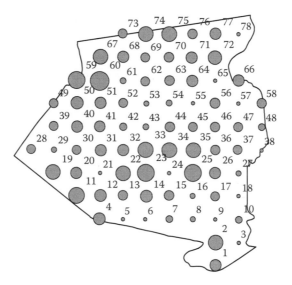

FIGURE 14.14
Percent of queen-move neighbors in same cluster as the locality, depicted using graduated symbols, with the smallest symbol denoting absence of like neighbors.

14.9 Trend Surfaces

Returning to Figure 14.7, there is strong visual suggestion of spatial trend surface behavior with regard to development. This is easily pursued in any of several software systems (Schabenberger and Gotway 2005). A second-degree polynomial trend surface model configured in **R** is as follows.

```
DvlTrnd <- lm(Dvl2k ~ PtX + PtY + PtXX + PtYY + PtXY)
```

A summary for this model is as follows.

```
Coefficients:
              Estimate   Std.Error  t value Pr(>|t|)
(Intercept) -4.064e+05  6.067e+04   -6.700 3.94e-09 ***
PtX          1.635e-01  3.488e-02    4.688 1.28e-05 ***
PtY          2.897e-01  4.489e-02    6.452 1.11e-08 ***
PtXX        -6.214e-08  9.661e-09   -6.432 1.21e-08 ***
PtYY        -7.122e-08  9.621e-09   -7.402 1.99e-10 ***
PtXY         1.624e-09  1.168e-08    0.139     0.89
---
Signif. codes:  0 '***' 0.001 '**' 0.01 '*' 0.05 '.' 0.1 ' ' 1

Residual standard error: 13.67 on 72 degrees of freedom
Multiple R-squared: 0.5751,     Adjusted R-squared: 0.5456
F-statistic: 19.49 on 5 and 72 DF,  p-value: 3.198e-12
```

While ordinary least squares (OLS) are problematic in a spatial context (Bivand et al. 2008), there is no need to elaborate the obvious in this situation. All terms except the cross-product are obviously predictive, and the cross-product is obviously not. Therefore, it is eliminated from the model with the following result.

```
Coefficients:
              Estimate Std. Error t value Pr(>|t|)
(Intercept) -4.118e+05  4.644e+04   -8.867 3.30e-13 ***
PtX          1.667e-01  2.571e-02    6.486 9.19e-09 ***
PtY          2.928e-01  3.867e-02    7.572 8.96e-11 ***
PtXX         6.211e-08  9.594e-09   -6.474 9.69e-09 ***
PtYY        -7.145e-08  9.420e-09   -7.585 8.47e-11 ***
---
Signif. codes:  0 '***' 0.001 '**' 0.01 '*' 0.05 '.' 0.1 ' ' 1

Residual standard error: 13.58 on 73 degrees of freedom
Multiple R-squared: 0.575,      Adjusted R-squared: 0.5517
F-statistic: 24.69 on 4 and 73 DF,  p-value: 6.024e-13
```

Fitted values, residuals, and unsigned residuals are then calculated for the reduced model and incorporated in a data frame for export to mapping software. A deciles map of fitted values is shown in Figure 14.15 for comparison with Figure 14.7.

Figure 14.16 shows a deciles map of signed residuals from the spatial trend surface, and Figure 14.17 shows deciles of unsigned residuals. While the spatial trend shown in Figure 14.15 is substantial and must be taken into account, it is also evident from Figures 14.16 and 14.17 that the residuals do not reflect second-order stationarity. Therefore, the simpler versions of geostatistical models will be problematic (Webster and Oliver 2001).

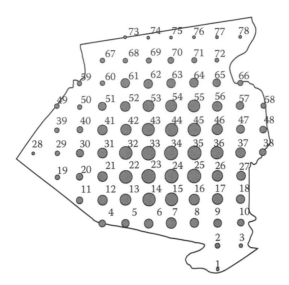

FIGURE 14.15
Deciles map of fitted trend surface values for percent development.

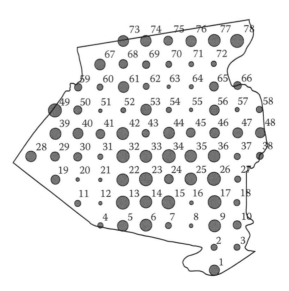

FIGURE 14.16
Deciles map of trend surface signed residuals for percent development.

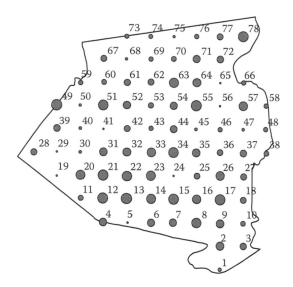

FIGURE 14.17
Deciles map of trend surface unsigned residuals for percent development.

14.10 Surveillance Systems: Sentinel Stations and Signaling

Localization of spatial data with IVIs provides a paradigm for conjunctive consideration of disparate data in both multivariate and univariate modes across several scales and with a variety of visualizations.

The localized layers have the overall character of a point-referenced, regular multivariate lattice. They constitute a sort of spatial metamodel that can couple information from different domains such as points, poly-lines, and polygons. With some adaptations, an array of analytical approaches can be applied from GIS, spatial statistics, partial order theory, and multivariate methods. GIS capabilities are needed to accomplish the localization, which can then be transferred to numerous other statistical software settings. Among these software settings, we have found the core and CRAN facilities of **R** to be powerful, economical, and readily available, especially when combined with algorithmic strategies structured in the open-source Python environment.

While localization provides a suitable structure for spatial surveillance with integrative indicators and ensuing etiology, a shift in perspective can help one to focus on function in the manner that object-oriented programming (OOP) of software shifts from procedural perspective to active analysis. We advocate a virtual view of simulated sentinels that are situated at stations from which they send signals regarding the status of the setting in which they are situated. The points of the locality layer are the *sentinel stations*, the proximity polygons of the localizing layer are the *sentinel settings*, and the vicinity variates are the *spatial signals*. What appear on the scene when viewed in this manner are the simulated sentinels themselves, which become organizing objects for analytical abilities as OOP methods and imbedded intelligence. Thus, the first step toward a functional focus would be to cast sentinels as classes of conceptual objects for OOP having methods of managing their computational context. The objects can then be called upon as needed to exercise their capabilities.

14.11 Scripted Sentinels

What distinguishes the sentinel situation from ordinary OOP is involvement and interplay in complex scenarios that are collective concerns such as with subsets of sentinels that may be comprised of clusters. Quasi-military modes make appropriate analogies whereby command center(s) control collectives of simulated sentinel entities by mounting missions. If command and control are by human hands, then it is likely that details of orchestrating recurring missions will be forgotten between sessions or that responsibility will be transferred from one analyst to another in the interim. Thus, scripts need to be prepared for each of the operational aspects and incorporated either formally or informally into the overall surveillance system.

Rudimentary regimes will have the scripts as detailed directions in a methods manual. Somewhat sophisticated systems will have decision support software (Sugumaran and Degroote 2011) and special software "shell scripts" (Robbins and Beebe 2005) that implement interfaces between software subsystems that participate in surveillance scenarios.

14.12 Smart-Sentinel Socialization

The future vision of soft-sentinel surveillance embraces the domains of agent-based software systems (Heppenstall et al. 2012; Luck et al. 2004) composed of "intelligent" cyber-entities with particular purposes and having capability for cooperating as collectives in higher levels of organization. Such cyber societies resemble colonies of super ants that orchestrate operations by collective control.

References

Abonyi, J., and Balaz, F. *Cluster Analysis for Data Mining and System Identification.* Berlin: Birkhauser, 2007.

Allerhand, M. *A Tiny Handbook of R.* New York: Springer, 2011.

Bivand, R., Pebesma, E., and Gomez-Rubio, V. *Applied Spatial Data Analysis with R.* New York: Springer, 2008.

Bruggemann, R., and Patil, G. *Ranking and Prioritization for Multi-Indicator Systems.* New York: Springer, 2011.

Burrough, P., and McDonnell, R. *Principles of Geographical Information Systems.* New York: Oxford Univ. Press, 1998.

Chander, G., Huang, C., Yang, L., Homer, C., and Larson, C. Developing consistent Landsat data sets for large area applications—the MRLC protocol. *IEEE Geoscience and Remote Sensing Letters* 6(4): 777–781, 2009.

Chrisman, N. *Exploring Geographic Information Systems.* New York: Wiley, 2002.

Costa, M., and Kulldorff, M. Applications of spatial scan statistics: A review. Chapter 6. In: Glaz, J., Pozdnyakov, V., and Wallenstein, S., Eds. *Scan Statistics: Methods and Applications.* Boston: Birkhauser, 2009.

Demers, M. *Fundamentals of Geographic Information Systems.* New York: Wiley, 2009.

Heppenstall, A., Crooks, A., See, L., and Batty, M., Eds. *Agent-Based Models of Geographical Systems.* New York: Springer, 2012.

Hogan, T. *Bare-Bones R: A Brief Introductory Guide.* Los Angeles: Sage, 2010.

Homer, C., Huang, C., Yang, L., Wylie, B., and Coan, M. Development of a 2001 National Landcover Database for the United States. *Photogrammetric Engineering and Remote Sensing* 70(7): 829–840, 2004.

Horton, N., and Kleinman, K. *Using R for Data Management, Statistical Analysis, and Graphics.* Boca Raton, FL: Taylor & Francis/CRC, 2011.

Kulldorff, M. A spatial scan statistic. *Communications in Statistics: Theory and Methods* 26: 1481–1496, 1997. [on-line]. Accessed August, 2011.

Lloyd, C. *Local Models for Spatial Analysis, 2nd ed.* Boca Raton, FL: CRC Press, 2011.

Luck, M., Ashri, R., and d'Inverno, M. *Agent-Based Software Development (Agent-Oriented Systems).* Norwood, MA: Artech House Publishers, 2004.

Lutz, M. *Learning Python.* Sebastopol, CA: O'Reilly Media, 2009.

Myers, W., and Patil, G. P. Biodiversity in the age of ecological indicators. *Acta Biotheoretica* 54: 119–123, 2006.

Myers, W., and Patil, G. Preliminary prioritization based on partial order theory and *R* software for compositional complexes in landscape ecology, with applications to restoration, remediation, and enhancement. *Environmental and Ecological Statistics* 17: 411–436, 2010.

Ormsby, T., Napolean, E., Burke, R., Groessl, C., and Bowden, L. *Getting to Know ArcGIS Desktop.* Redlands, CA: ESRI Press, 2010.

Myers, W., and Patil, G. P. Geoinformatics for Human Environment Interface. Proc. Joint Stat. Meetings (JSM) 2011, July 31, 2011, Miami Beach, FL, session 206322, presentation 300319, www.amstat.org on-line archives.

R Development Core Team. *R: A Language and Environment for Statistical Computing.* Vienna, Austria: R Foundation for Statistical Computing. ISBN 3-900051-07-0, URL http://www.R-project.org/, 2008.

Robbins, A., and Beebe, N. *Class Shell Scripting*. Sebastopol, CA: O'Reilly Media, 2005.

Schabenberger, O., and Gotway, C. *Statistical Methods for Spatial Data Analysis*. Boca Raton, FL: Chapman & Hall/CRC, 2005.

Sugumaran, R., and Degroote, J. *Spatial Decision Support Systems: Principles and Practices*. Boca Raton, FL: CRC Press, 2011.

Summerfield, M. *Programming in Python 3: A Complete Introduction to the Python Language*. Upper Saddle River, NJ: Addison-Wesley Professional, 2009.

Venables, W., Smith, D., and the R Development Core Team. *An Introduction to R*. Bristol, UK: Network Theory Ltd., 2005.

Wallenstein, S. Joseph Naus: Father of the scan statistic. Chapter 1. In: Glaz, J., Pozdnyakov, V., and Wallenstein, S., Eds. *Scan Statistics: Methods and Applications*. Boston: Birkhauser, 2009.

Webster, R., and Oliver, M. *Geostatistics for Environmental Scientists*. New York: Wiley. 2001.

Xu, R., and Wunsch, D. *Clustering*. New York: Wiley, 2009.

Index

A

Age of indicators, 2
Allegheny County, 191–197, 201
American Standard Code for
 Information Interchange
 (ASCII), 19, 44
ArcGIS, 19, 21, 28, 113–114
 land cover data, 50
ArcGIS Explorer, 30
Associated attributes, 51–52
Automathics, 3

B

Buffer geometries, 19–20

C

Cartography, 47–48
Census tracts and blocks, 12
Centroids
 CLAN cluster, 68–69
 cluster, 200
 salient, 69–71
CIT core development
 complexion of, 100–104
 hotspot of, 97–100
Cluster companions, 144–147, 200–202
Clustered localities agglomerated
 nonspatially (CLANs), 57
 cluster centroids, 68–69
 clusters, 59–63, 144
Clustered localities using map positions
 (CLUMPs), 57, 64–68
Coloring maps, 47–48
Command-based statistical software
 system, 44
Comprehensive R Archive Network
 (CRAN), 44, 171
Cultural conditioning of ecology, 2
Cyber societies, 206

D

Decile map
 of local spatial association, 152
 for percentage of developed land, 177
 for percentage of forested land, 177
Dendrogram of clustering, 60
 for CLUMP components, 65–66
 spatial clustering, 64
Digital raster graphics (DRG), 8
Distal data, 129–132

E

Earth Observation Systems (EOS)
 programs, 12
Environmental Systems Research
 Institute (ESRI), 3
*Environomics of Environmentally Safe
 Prosperity,* 2

F

Federal Information Processing
 Standards (FIPS), 12
FGIS facility, 30

G

Geary's *C,* 160
Generalized land cover layer, 55
Generic GIS, 28–30
GeoDa, 165
Geodata, 155
GeoDa™ software, 159
Geographically weighted regression
 (GWR) approaches, 171
Geographic Resources Analysis Support
 System (GRASS) GIS, 30
Geoinformatics, 3–4
 localization as vicinity variates, 4–5
 spatial synthesis, 4–5

GeoR, 155
Georeferencing, 3, 12
Geospatial ecoinformatics, 2
Geospatial One Stop (GOS), 12
GIS software, 43–45
    vector data, 47

H

Hotspot, 95–96
    of CIT core development, 97–100
    space–time, 183
H-scatterplot, 154–155
Human interface
    census information, 12
    distributed data depots and digital
        delivery, 12–13
    environment information, 1–3
    localization as vicinity variates, 4–5
    provisional proximity perimeters
        and pattern of posting points,
        7–10
    Spatial Posting of Tabulations
        (SPOTing), 5–6
    spatial synthesis, 4–5
    sustainability of environment, 2
    in terms of counties, 6–7

I

IDLE, 110
Image algebra, 80–81
Inconsistency indicator, 42–43
Integrated vicinity indicators (IVIs), 5,
    21, 31, 79, 113
    vs cluster number, 62–63
Intensity image information
    hillshading and slopes, 86–88
    intensity as frequency of occurrence,
        79–86
    interposed distance indicators (IDIs),
        79, 88–89
    perceptions of pictures, 90–93
    pixel-by-pixel information, 90–93
    spectral selectivity and sensitivity, 91
    of Wyoming Valley, 91–92
Intensity indicator, 177–180
Interpolation, 177
    error, 181

inverse distance weighting (IDW),
    178–179
    by kriging, 181–182
    for percent forest, 178
    spline, 180
Interposed distance indicators (IDIs), 5,
    79, 88–89
Inverse distance weighting (IDW),
    177–179

K

Kindred clusters, 147
Kriging, 181–182

L

Lackawanna County
    decile diagram, 54
    land cover data, 50–51
    map of ID numbers for localities, 61
    octagons spanning partial polygons,
        123–124
    point positions, 174–175
    provisional proximity perimeters and
        pattern of posting points, 7–10
    scatterplots for IVIs and their IDs
        for, 32
Land cover
    classification, 48–51
    intensity image information, 79–80
Land Processes Distributed Active
        Archive Center (LPDAAC), 12
Lattice plot facility, 32
Layer logic, 55
Local indicator of spatial association
    (LISA), 151–152
Localization paradigm
    apportioning attributes of partial
        polygons, 28
    with densities, 22–24
    of land cover data, 56
    locality layer, 193–194
    localizing layer, 191–195
    as poly-place purview, 15–18
    of proximity perimeters, 19–22
    using Excel, 26–28
    using generic GIS, 28–30
    as vicinity variates, 4–5

Low-density residential land, mapping of, 79–80
   LoddIVI indicator for, 126
Low-order polynomial trend surface, 162

M

Major maximal rank, 129
Map algebra, 55
   commercial/industrial/
      transportation, criteria for,
      82–84
   residential development, criteria for, 82
Map modeling, 52–55
MapWindow©, 19, 29–30
Matrix object, 44
Metadata, 12
Micromapping, 47–48
Micromaps, 189
Minor maximal rank, 129
Montreal Process, 2
Moran's *I*, 150, 158–160, 167
Multi-Resolution Land Characteristics
      (MRLC) consortium, 48
MultiSpec©, 91

N

National Land Cover Database
      (NLCD), 48
National Spatial Data Infrastructure
      (NSDI), 12
Nugget effect, 157

O

Object-oriented (OOP) program, 110, 205
Octagonal integration vicinities
      (OCTIVs), 20
Octagonal proximity perimeters, 20
Octagonal vicinities (OCTIVs), 175
Octagons, 113–117
   generation using Python (2.7)
      program, 114–116, 118
   matching margins and adjusting
      areas, 117–119
   shape and support for local roads,
      119–121
   spanning partial polygons, 123–124

Open Geospatial Consortium (OGC), 30
ORDering In Tandem (ORDIT), 38–39, 70
Ordinary least squares (OLS), 203
   method of estimation, 165

P

Paired plots, 186
Pairs at lagged locations, 152–155
PAMAP (PA MAPping) program of
      digital aerial photography, 90
Pearson correlations, 32
   matrix, 57
Pennsylvania Spatial Data Access
      (PASDA) portal, 12
Planar plotting, 13
Polygons, 113
   octagons spanning partial, 123–124
Poly-place purview, 196–197
Precedence, approach to, 34
   partial (percent) pairwise
      precedence, 198–200
   plot for shapes and supports, 121–123
   plot with special annotation, 36–38
   product–order protocol, 35–36
   propensities as progression of, 38–40
   signals for plots, 125–128
Primary partition plots, 186–188
Product–order protocol, 199
Progression plot, 40–41
Proximity perimeters
   for Lackawanna County, 21
   localization, 19–22
   octagonal, 117–119
   referencing of, 7–10
   transfer of the IVI data to posting
      points, 24–28
   using Python (2.7) program, 104–109
   vicinity indicators, 175–176
   XurbnIVI for, 84–85
Python program, 104–110
   averages for rings of distance, 147–150
   count cluster companions, 144–147
   to generate grid of points, 173–174
   for generating octagons from center
      points, 114–116, 118
   for a pairing approach, 153–154
Python Spatial Analysis Library
      (PySAL), 110

R

Race ranking, 41–42
Range, 157
Rank correlation matrix, 128–132
Rank rods, 72–74
Raster representation, 51–52
R base system, 44
Remote sensing science, 90–93
    spectral detection of change with,
        189
Representative ranks, 71–72
    salient sequences, 74–78
Restricted regression, 171
R function, 34–35, 110, 132
    built-in ranking facility in, 41
    for collecting ranks and extracting
        representatives, 71–72
    ORDITing function, 39–40
    for plotting range rods, 73–74
    product–order precedence function
        for, 35–36
R© statistical software system, 31

S

Salient sequence
    of representative ranks, 74–78
    scatterplot of, 183–186
SaTScan™, 96–97, 100, 103–105, 183, 197
Scale sensitivity, 198–200
Scripted sentinels, 206
Scripting language, 110
Second-degree polynomial trend
        surface model, 161–162
Semisynchronous signals, 125
    concept of pivot positions and
        pairing positions, 140
    correlation matrix for
        synchronization shifts, 133, 135
    CrosCnts function, 138
    CrosCpld function facility, 139
    CrosRank function, 136–137
    CrossNdx function, 137–138
    of distal data, 129–132
    LoddIVI, 127–128, 140
    MdnMdlSS function, 132–133

median models of signal structure,
        132–136
    pairing/placement patterns,
        136–141
    parallel boxplots of the distal data,
        130
    related to CLAN and CLUMP
        community clusters, 141
Semivariogram value, 155
Sentinel settings, 10–11, 205–206
    stations and signaling, 205–206
Shell scripts, 206
Sill, 157
Smart-sentinel socialization, 206
Soft sentinel system, 10–11
Space–time hotspot, 183
Spatial autocorrelation, 143, 158
Spatial data clearinghouses, 12
Spatial error model, 165, 167
Spatial lag model, 164, 169
Spatial Posting of Tabulations
        (SPOTing), 5–6
Spatial regression model, 171
Spatial signals, 206
Spearman correlations, 32, 57–58
Spline smoothing, 180–181
Spline strategies, 180–181
*Statistical Geoinformatics for Human
        Environment Interface*, 1
Statistical software, 43–45
Surveillance systems, 205–206
Sustainability of environment, 2

T

Topologically Integrated Geographic
        Encoding and Referencing
        (TIGER) data, 12
Trend surface model, 161–164, 202–205
Trigonometric analysis, 113

U

University Consortium on Geographic
        Information Sciences
        (UCGIS), 4

V

Value attribute table (VAT), 52
Variograms, 155–158, 181
Vicinity variates
    localization as, 4–5
    rank correlations for, 128
    regression relations among, 164–170

Virtual topography, 86
Visualizations, 47–48

W

Wyoming Valley, 157, 174, 180

Printed and bound by CPI Group (UK) Ltd, Croydon, CR0 4YY

21/10/2024

01777089-0003